Ancient Jewish Sciences and the History of Knowledge in Second Temple Literature

Ancient Jewish Sciences and the History of Knowledge in Second Temple Literature

Editors

Jonathan Ben-Dov

and

Seth Sanders

New York University Press
and
Institute for the Study of the Ancient World
2014

Gentium Plus and EZRA fonts provided by SIL International and are used under terms of the Open Font License.

At the time of publication, the full-text of this work was available at:
http://dlib.nyu.edu/awdl/isaw/ancient-jewish-sciences/ .

Library of Congress Cataloging-in-Publication Data

Ancient Jewish sciences and the history of knowledge in Second Temple literature / editors Jonathan Ben-Dov and Seth Sanders.
volumes cm
Includes bibliographical references and index.
ISBN 978-1-4798-2304-8 (cloth) -- ISBN 978-1-4798-7397-5 (ebook) -- ISBN 978-1-4798-6398-3
1. Science, Ancient--History. 2. Astronomy, Ancient. 3. Astrology, Hebrew. 4. Judaism and science. 5. Physiognomy--Religious aspects--Judaism. I. Ben-Dov, Jonathan, editor. II. Sanders, Seth L., editor.
Q124.95.A53 2013
509.33'09014--dc23
2013016449

ISBN 978-1-47982-304-8 (cloth)
ISBN 978-1-47987-397-5 (ebook)
ISBN 978-1-47986-398-3 (ebook)

Contents

Acknowledgments

Jonathan Ben-Dov and Seth L. Sanders

This book was an unexpected positive side effect of our fellowship at the Institute for the Study of the Ancient World in New York University during 2010/11. What started as a small adventure turned out to be a meaningful conference—both for us and we hope also for the attendees and the readers of the present book. We are especially glad to present the volume in the ISAW series through NYU Press, both as a printed book and as an electronic document, with free access to the public via Creative Commons.

It is a pleasant duty to acknowledge the help we received from people at ISAW, who did not outright reject our idea for a spontaneous conference but rather embraced it with enthusiasm. ISAW director, Roger Bagnall, accepted the conference to ISAW's schedule and supported the preparations in many ways. Alexander Jones lent a helpful hand and priceless advice in the organizing the conference, as well as delivering a response to one of the sessions. His presence at the lecture hall gave us the much desired perspective of the 'general' history of science outside the Jewish sources. Kate Lawson from ISAW was enormously helpful in putting the conference together, never tiring of our strange requests and special needs. Sebastian Heath has been an alert and graceful editor, who has a great impact on the book. Last but not least, we thank Shelby White, director of the Leon Levy Foundation, for her support of and engagement with the conference.

We thank John Collins, Seth Schwartz, and Lawrence Schiffman for chairing sessions and leading lively discussions at the conference. The issues each raised have had a significant impact in the pages of this book. Irene Soto, a graduate student from ISAW, quickly did most of the copy-editing, for which we are greatly indebted to her. Ross Teasler was instrumental in preparing the index.

Philip Alexander's 2002 article is reprinted here courtesy of Peeters Press (Leuven). We are grateful for both the author and the press for permission to include the paper in the present volume.

Jonathan Ben-Dov acknowledges the help of the Israel Science Foundation, as well as the kind hospitality and scholarly discussion on the part of the following people during 2010/11: Dan Machiela, Steven Fraade, Max Grossman, Hayim Lapin, and Eileen Schuller. Seth Sanders thanks Trinity College for research support and Joseph Angel, Michael Barany, Jacco Dieleman and Mathieu Ossendrijver for enlightenment.

1. Introduction

Jonathan Ben-Dov and Seth L. Sanders

1. *The Idea of Ancient Jewish Science*

Sometime after the end of the Judahite monarchy, Jewish writers opened their eyes to the universe in an unprecedented way.[1] A new interest in the cosmos and its patterns appears in late-Persian and Hellenistic apocalyptic literature. For the first time in Jewish literature, we find astronomy and cosmic geography—secrets lying beyond the traditionally understood and immediately visible world—in the *Astronomical Book* of Enoch and the *Book of Watchers*. Texts like the Aramaic Levi document and the Qumran physiognomies extend these interests from the stars to the measurement of materials and the human body. In these sources we find precise new ways to divide up and understand the world. The knowledge they present is of a sort unprecedented in Jewish sources because it contains detailed, systematic rules and observations about the physical world—what scholars of Greece and Babylon have long studied as ancient science.

But how did a type of science emerge in early Judaism when the Bible shows no interest in it, apparently prohibiting even inquiry into the stars (Deut. 4:19)? Why does this new Jewish science appear in such complex forms, intertwined with stories of biblical patriarchs and

[1] The present book and hence also this introduction address the systematic representation of scientific themes in literature from the Persian and Hellenistic periods. By doing this we leave aside the discussion of the themes of Nature and Creation in the Hebrew Bible. While mid-twentieth century scholarship downplayed the intensity of this involvement, highlighting instead the idea of divine involvement in history, it is now clear that some biblical authors entertained a comprehensive *Weltbild*, including an interest in the regularities and irregularities of nature. See *Das biblische Weltbild und seine altorientalischen Kontexte* (ed. B. Janowski and B. Ego; FAT 32; Tübingen: Mohr Siebeck, 2001); J. Ben-Dov, "Is there a Worldview in the Hebrew Bible?," *Shnaton* 15 (2005): 297-307 (Hebrew). Baruch Halpern argues for the existence of explicit cosmological paradigms in the Hebrew Bible resembling pre-Socratic cosmologies in *idem*, "The Assyrian Astronomy of Genesis 1 and the Birth of Milesian Philosophy," in *From Gods to God: the Dynamics of Iron Age Cosmologies* (FAT 63; Tübingen: Mohr Siebeck, 2009) 427-442.

"irrational" religious elements such as mystical visions? Who were these early searchers after knowledge, and what can we learn from the distribution of the earliest evidence, which is known only from the Qumran caves? Were the sectarians merely scribes copying traditional knowledge by rote or researchers doing experiments and formulating laws? Were their sources Babylonian, Aramean or Jewish—or is this distinction ahistorical? And what was the fate of their knowledge—does it ultimately dissolve into apocalyptic fantasy, or is it the beginning of a long and complex relationship between Judaism and science?

In this book, we outline this remarkable new kind of ancient knowledge. In addition to attempting to explain its rise, we also investigate the parameters of ancient Jewish science: the ways it might—and might not—be usefully understood as 'science;' how it might be understood as 'Jewish:' whether as a somehow inherently Jewish phenomenon, or simply science practiced by Jews.[2]

This book is the first to bring major scholars together to explore the relationship between science and early Judaism. It addresses a set of essential problems which traditional scholarship has rarely recognized and is perhaps not well suited to address. So it is fitting that it emerged from a conversation in the hallway of the Institute for the Study of the Ancient World, an institution dedicated to crucial but sometimes invisible interconnections, not only between ancient cultures but between modern disciplines.

If this book's specific questions are new, its larger issues are not. The work presented here rests on the foundations of an established discipline, which has studied the role of Jews in the formation of medieval and early modern science. Beginning with Amos Funkenstein's pioneering studies and continuing with Ruderman's now classic *Jewish Thought and Scientific Discovery in Early Modern Europe*, it is by now an established fact that Jews not only took part in scientific activity in these periods but felt compelled to provide a theological basis for that activity.[3] Following the establishment of this

[2] See the reasoning in the landmark book by Murray J. Rosman, *How Jewish is Jewish History?* (Oxford: Littman Library of Jewish Civilization, 2007).

[3] Amos Funkenstein, *Theology and the Scientific Imagination from the Middle Ages to the Seventeenth Century* (Princeton: Princeton University Press, 1986);

discipline for the early modern period, it was carried further by scholars of medieval Jewish literature.[4] This branch of historiography draws attention to the blurred borderlines, and often actual continuity, between what we would call 'science' and other disciplines often classified as esoteric, mainly Kabbalah. It is one of the aims of the present volume to carry this research agenda forward. Not surprisingly, the conflicted boundaries of science and esotericism constitute a central focus of interest in the ancient material too. This connection with secret knowledge is especially clear in our case, because the scientific trend in early Judaism arose in the framework of apocalyptic groups and their reflection on world order. But in fact, the interplay between science and esotericism has been as relevant to non-Jewish scientists as to their Jewish contemporaries throughout the historical periods mentioned above.[5]

A clarification is due about the title chosen for this book: *Ancient Jewish Sciences*. The studies mentioned above focus on Jewish *scientists* rather than on Jewish *science*. Modern science is typically understood as a universal and objective venture, with no meaningful distinction between Jewish and Christian, European or Asian science. Indeed, the term 'Jewish Science' appeared as an allegation against Jews during the Third Reich, to undermine their role in science as ethnically distinct, and therefore suspect.[6]

David B. Ruderman, *Jewish Thought and Scientific Discovery in Early Modern Europe* (New Haven and London: Yale University Press, 1995).

[4] Gad Freudenthal, *Science in Medieval Jewish Cultures* (New York: Cambridge University Press, 2011); Y.T. Langermann, *The Jews and the Sciences in the Middle Ages* (New York: Ashgate, 1999); S. Sela, *Abraham Ibn Ezra and the Rise of Medieval Hebrew Science* (Leiden: Brill, 2003).

[5] See for example A. Yarbro-Collins, *Cosmology and Eschatology in Jewish and Christian Apocalypticism* (JSJSup 50; Leiden: Brill, 1996).

[6] See Dirk Rupnow, *Judenforschung im dritten Reich. Wissenschaft zwischen Politik, Propaganda und Ideologie* (Baden-Baden: Nomos, 2011). Thanks are due to Amos Morris-Reich and to Dirk Rupnow for this reference.

Earlier, a spiritualist movement called 'Jewish Science' arose as a creative response to the problems of secularizing American Jewish life in the early 20th century. On this movement see Ellen Umansky, *From Christian Science to Jewish Science: Spiritual Healing and American Jews* (Oxford: Oxford University Press, 2005). Needless to say, while this volume's title has some remote and complex historical resonance with both of these dramatically different 20th-century terms, there is no direct connection.

But things were different in ancient times, when the Jewish nature of a science could be part of the point. In the Hellenistic period, when the search for indigenous *protos heuretes* ("first discoverers") was at its height, some practitioners of ancient science declared themselves to be searching for a specifically Jewish contribution to the field of cosmology and science, as attested for example in Pseudo-Eupolemus' famous claim that Enoch "discovered" astronomy (engaged below in the articles by Sanders and Reed). Thus the title "Ancient Jewish Sciences" fits quite loyally with the spirit of our objects of research. More profoundly, there is something inherently Jewish in the disciplines studied here, which depart from the practice of, say, Jewish medieval scholars like Abraham Ibn-Ezra. The modes of production and articulation of scientific material by ancient Jewish apocalyptic groups were often specifically geared to the needs of the community and to its theological worldview. Thus, as Ben-Dov points out in his article, the organizational needs of the Yahad required the development of a special branch of science, hitherto unattested in other ancient literature. In this way, therefore, the subject-matter of the present volume deserves to be called 'Jewish Science'.

In the present volume we seek to present a more intricate view of the tension between the universality of scientific knowledge and the uniqueness of local traditions. This tension has been restudied in recent decades within the discipline of the history of science.[7] It is increasingly clear, first, that scientific traditions develop in non-modern societies according to the unique circumstances of each and every one of them. Indeed, this has also been demonstrated for the politically and economically central aspects of "hard" sciences in modern societies such as nuclear physics and petroleum geology.[8]

[7] See especially the theme issue "Nature and Empire. Science and the Colonial Enterprise," edited by Roy MacLeod, *Osiris* n.s. 15 (2000).

[8] Probably the most direct impact of physics on 20th-century politics was the creation of the controlled atomic fission reactions that destroyed Hiroshima and Nagasaki, helping end World War II and lay the grounds for the "Cold War" that defined the next four decades of geopolitics. Concern over the military value of this science resulted in the remarkable approach to knowledge signed into law as the 1946 Atomic Energy Act, which the eminent historian of science Peter Galison has described as an "anti-epistemology." As Galison puts it, "Nuclear weapons knowledge is born secret ... [it] becomes classified the instant it is written down." P.

Second, each of these local traditions maintains a mutual relationship with the prevalent scientific culture, at once absorbing elements from and leaving its mark on it. It is thus enlightening to examine a peripheral manifestation of scientific culture in antiquity according to these newly available tools. This methodological aspect proves to be especially relevant with respect to the ancient Jewish material, due to its unique location in space and time; it lies between the great centers of Mesopotamia and Hellenistic Egypt, at a time when the encounter between cuneiform science and the West is in full swing. The Jewish material lies in time between Hipparchus and Ptolemy, who both used cuneiform data in various ways, while its contents float somewhere between cuneiform and Greek scientific traditions.[9] The formation of a

Galison, "Removing Knowledge" *Critical Inquiry* 31 (2004): 232. Howard Morland describes the remarkable implications of this law: "When the Atomic Energy Act became law, it defined a new legal term 'restricted data' as 'all data concerning the manufacture or utilization of atomic weapons, the production of fissionable material, or the use of fissionable material in the production of power,' unless the information has been declassified. The phrase 'all data' included every suggestion, speculation, scenario, or rumor —past, present, or future, regardless of its source, or even of its accuracy— unless it was declassified. All such data were born secret and belonged to the government. [In terms of the original Act, i]f you related a dream about nuclear weapons, you were breaking the law." —Morland, "Born Secret," *Cardozo Law Review* 25 (2005): 1402. Even after significant modification, this essential anti-epistemology, in which types of nuclear knowledge are inherently classified, is retained in U.S. Law: "a positive action is not required to put information into the [Restricted Data] category. If information falls within the Act's definition of RD, it is in this category from the moment of its origination; that is, it is 'born classified'," as classifier Arvin Quist wrote in a 2002 report commissioned by the Oak Ridge National Laboratory, *Security Classification of Information,* 2 vols., http://www.fas.org/sgp/library/quist/index.html, vol. 1, p. 88. For the Nuclear Regulatory Commission's current definition of inherently classified "Restricted Data," and the rules by which it can be declassified, see sections 11 and 142 of the Atomic Energy Act of 1954 (P.L. 83-703) at http://www.nrc.gov/about-nrc/governing-laws.html#aea-1954. Similarly, in the oil industry, perhaps the most powerful business in the world, significant geological research is both created and concealed from the public because it concerns whether there is oil in a given area and is thus the trade secret of the company that sponsored it. For petrogeological secrets see Timothy Mitchell, *Carbon Democracy: Political Power in the Age of Oil* (New York: Verso, 2011), 244 and for a case study Geoffrey Bowker, *Science on the Run: Information Management and Industrial Geophysics at Schlumberger, 1920-1940* (Cambridge, MA: MIT Press, 1994).

[9] See especially Jonathan Ben-Dov, *Head of all Years: Astronomy and Calendars at Qumran in their Ancient Context* (STDJ 78; Leiden: Brill, 2008) and Mladen

national narrative of scientific discovery in this context constitutes a particularly valuable case study which we aim to pursue here.

In what follows, we will introduce a few of the most interesting problems that the scientific elements in Ancient Judaism present, and the essays in this volume that contribute to solving them.

2. Why "Science"? The Demarcation Problem and the Danger of Reinventing the Wheel

In the Second Temple period, remarkable new types of knowledge and genres of text appear in Jewish culture. These include astronomical calculations of the movements of the heavenly bodies and length of the days, sexagesimal (base-60) metrology, simple forms of zodiacal astrology, and physiognomic interpretation of the body. They systematically describe aspects of the physical world in a precise new way—usually a way first developed in Mesopotamia. And all these modes of knowledge have at some point in modern European history been understood as natural science: astronomy and mathematics are of course still understood this way, but as late as the mid-19th-century a form of physiognomy known as "phrenology" was taken seriously by scholars across Europe.[10]

But is it science? It seems intuitively correct to us to define mathematics and astronomy as exact science, but is it science to observe someone's hair to predict their character and destiny, as the Qumran physiognomic text 4Q186 does?[11] As Sanders shows in his essay, the history and philosophy of science provide a surprising but clear answer: there is no rigorous way to tell.

Popović, *Reading the Human Body: Physiognomics and Astrology in the Dead Sea Scrolls and Hellenistic-Early Roman Period Judaism* (STDJ 67; Leiden: Brill, 2007).

[10] On the lives and deaths of phrenology and related physical and quantitative approaches to human character, see the lively study of Stephen Jay Gould, *The Mismeasure of Man* (New York: Norton, 1996, Rev. and expanded ed).

[11] An illustrative passage comes from 4Q186 1 ii 5-8, which we translate: "[And anyone] whose thighs are long and slender, whose toes are slender and long, and he is from the second column: he possesses a spirit with six parts light, but three parts in the House of Darkness. This is the birth sign (horoscope) under which he was born: the foot of Taurus. He will be humble/poor. This is his animal: the bull."

In an influential 1983 article, the philosopher of science Larry Laudan explained that the problem of distinguishing scientific knowledge from other types has loomed large in Western philosophy for a long time:

> From Plato to Popper, philosophers have sought to identify those epistemic features which mark off science from other sorts of belief and activity.[12]

In the philosophy of science, the task of defining the boundaries of science became known as the "demarcation problem," and after well over a century of heated debate it is now generally agreed to be insoluble:

> ... it is probably fair to say that there is no demarcation line between science and non-science, or between science and pseudo-science, which would win assent from a majority of philosophers (112).

It appears that historically, no necessary and sufficient definition of science (or its ancestors such as Aristotle's *episteme*) has ever been devised. As Laudan explains, definitions have at some points focused on science as proceeding deductively from *a priori* principles and at other points as proceeding inductively from observed phenomena. Science has also been defined as being "falsifiable" (a definition that includes the "falsified" flat earth theory) or as proceeding from a "scientific method", the rules of which were never successfully explained. And as Francesca Rochberg has shown, ancient science has also been defined in contradictory ways. It has been described as inhering in explanation without accurate observation (Greek), or accurate observation without explanation (Mesopotamia and Egypt).[13]

But this very debate over the nature of ancient science suggests a more promising avenue of inquiry. While there has been a tremendous amount of successful science done in the modern world, most

[12] "The Demise of the Demarcation Problem" in *Physics, Philosophy, and Psychoanalysis: Essays in Honor of Adolf Grünbaum* (ed. R.S. Cohan and L. Laudan; Boston Studies in the Philosophy of Science 76; Dordrecht: D. Reidel, 1983), 111.

[13] *The Heavenly Writing: Divination, Horoscopy, and Astronomy in Mesopotamian Culture* (Cambridge, UK & New York, NY: Cambridge University Press, 2004).

philosophers of science now agree that "science" itself is not something we can clearly and rigorously define. Most influentially, Larry Laudan showed that nobody has yet been able to find a set of conditions that is both necessary and sufficient to characterize everything we now consider "science." Instead, Laudan emphasizes "strong confirmation"—a standard of proof that may apply to disciplines like history or philology as well.

When scholars of early Judaism improvise definitions of science, we freshly encounter an old problem now usually considered by philosophers and historians of science to be insoluble because its object is so heterogeneous. As Laudan writes, "... it may just be that there are no epistemic features which all and only the disciplines we accept as 'scientific' share in common." The historian of science Steven Shapin concludes, "You could say that science is not one, indivisible, and unified, but that *the sciences* are many, diverse, and disunified."[14]

Awareness of this long debate in the philosophy of science can save us from reinventing the wheel: if neither Karl Popper nor Imre Lakatos could come up with a solid, broadly applicable definition of science, we should not be embarrassed if our own attempts come to grief as well.[15] And this awareness may help us shift our focus to the question of what we seek to learn. If we do invent a wheel, it should help take us someplace we want to go.

If we find the category of "science" to be a useful one for early Judaism, a second question arises, this one concerning its historical emergence. Would the appearance of such a pattern represent a rupture from long-term Judean/Jewish/Hebrew[16] discourses and

[14] Laudan, "Demise" 112; Steven Shapin, *Never Pure: Historical Studies of Science as if it was Produced by People with Bodies, Situated in Time, Space, Culture, and Society, and Struggling for Credibility and Authority* (Baltimore, Md.: Johns Hopkins University Press, 2010), 5.

[15] For Lakatos' theory of scientific progress and critiques of Kuhn and Feyerabend, see Larry Laudan, "Reconciling Progress and Loss," in *Beyond Positivism and Relativism: Theory, Method, and Evidence* (Boulder: Westview, 1996), 113-122.

[16] *n.b.* not a smoothly overlapping series—surely part of the point, and perhaps part of the solution? As Michael Stone points out in his response to Ben-Dov's article, "Émile Puech [argues that] the linguistic milieu of Qumran was no different from that of the rest of contemporary Judea. I

traditions? If there are no biblical genres of mathematics or astronomy, for example, and divination is explicitly taboo, the extensive use of Babylonian astral science in Qumran Aramaic texts would represent a radical break.[17] This would foreground the problem of historical change: what forces led to this break? Should we see it emerging from a cosmopolitan Aramaic world of exact descriptive knowledge in opposition to or initial isolation from Priestly and Deuteronomistic categorizations?[18] If, on the other hand, Jewish science is bound up with scriptural exegesis, this would draw attention to emergent "scriptural" Hebrew texts like Genesis 1-2:4a, Exodus 25-31 (cf. Ex 35-40 and Ezekiel 40-48), and Leviticus 12-15, with their exact descriptions of cosmos, temple, and human phenomena (and in Leviticus, with the explicit command to closely observe physical signs to "diagnose" them as "symptoms," primarily of *ṣara'at*). Again, do

propose considering that the same is true of the 'scientific' milieu. In fact, we have very little information about the greater culture in which the Jews in the land of Israel lived, either in the First or Second Temple periods. If we were dependent on the Hebrew Bible, virtually nothing, for the Hebrew Bible does not deal with scientific issues ... All considered, however, it is probable that the "larger culture" in which the Jews lived was basically Mesopotamian." Stone *apud* Ben-Dov "Scientific Writings in Aramaic and Hebrew at Qumran: Translation and Concealment" in *Aramaica Qumranica: Proceedings of the Conference on the Aramaic Texts from Qumran in Aix-en-Provence 30 June - 2 July 2008* (eds. K. Bertholet and D. Stoekl Ben Ezra; STDJ 94; Leiden: Brill, 2010), 400.

[17] For Deuteronomistic prohibitions of divination see Dt 18:10, 14; I Sam 15:23; 2 Kings 17:17 and for further discussion see the chapter by Sanders in this volume. By contrast, study of the heavenly bodies is presented as an impetus to both obedience and disobedience (1 En 2) and blasphemy (1 En 7-9) in the editorially complex *Book of Watchers*. Annette Reed explores this tension in *Fallen angels and the History of Judaism and Christianity: The Reception of Enochic Literature* (New York: Cambridge University Press, 2005), 37-44.

[18] Baruch Halpern has argued for the influence of Assyrian astronomy and cosmology already in the Priestly source of Genesis 1-2:4a; for a brief critique see Sanders in this volume. For the intertwining of the Babylonian and Aramaic script-languages and intellectual worlds see the rich presentation of Paul-Alain Beaulieu, "Official and Vernacular Languages: The Shifting Sands of Imperial and Cultural Identities in First-Millennium B.C. Mesopotamia" in *Margins of Writing, Origins of Cultures* (ed. S. L. Sanders; University of Chicago Oriental Institute Seminars 2; Chicago: Oriental Institute, 2006), 187–216.

these texts belong on the far side of a pre-science/science divide?[19] And if so how was it bridged?

3: Science between Local Tradition and the Discovery of Universals

Sanders and Ben-Dov intended the conference as the beginning of a public conversation. Held April 4, 2011 at ISAW in New York,[20] the conference began by addressing the most fundamental evidence through an in-depth discussion of the *Astronomical Book* of Enoch, the earliest known Jewish—and Aramaic—scientific work, and closely related texts. The second half of the conference worked outward to the earliest Jewish communities in which science could have been conducted and concluded with a wide-ranging discussion of the stakes of understanding these early Jewish activities as scientific practice.

The rest of this introduction will sketch the contributions and suggest their possible significance. The *Astronomical Book* of Enoch, an originally independent Aramaic treatise better known from its present position in chapters 72-82 of 1 Enoch, is the subject of discussion between the first four contributors, Philip Alexander, James VanderKam, Seth Sanders and Loren Stuckenbruck. They apply a variety of methods to this core text of the scientific—and apocalyptic—tradition, thus making the present book an outstanding laboratory of attitudes for dealing with a single proof text.

An article by Philip Alexander from 2002 is reprinted here because it in many ways set the stage for the present discussion by contextualizing the *Astronomical Book* in the study of ancient science. As is often the case with pioneering studies, Alexander's essay created

[19] Michael Stone noted the continuities with geographical and cosmographical lists in his "Lists of Revealed Things in the Apocalyptic Literature" in *Magnalia Dei, the Mighty Acts of God: Essays on the Bible and Archaeology in Memory of G. Ernest Wright* (ed. Frank Moore Cross, Werner Lemke, and Patrick D. Miller; Garden City N.Y.: Doubleday, 1976, 1st ed.), 414-452; for further discussion see Sanders' essay in this volume.

[20] The original lineup was: James VanderKam (University of Notre Dame) and Seth Sanders (Trinity College and ISAW), with Loren Stuckenbruck (Princeton Theological Seminary) as respondent; Jonathan Ben-Dov (University of Haifa and ISAW) and Mladen Popović (University of Groningen), with Alexander Jones (ISAW) as respondent; and Annette Yoshiko Reed (University of Pennsylvania), with Lawrence Schiffman (NYU) as final respondent. John Collins (Yale) and Seth Schwartz (Columbia) chaired the two sessions.

a point of departure for later authors. Even detailed critiques (see e.g. Reed, this volume) tend to uphold the general framework it established. Alexander attempted to outline a distinct Jewish tradition, which began already in biblical literature and continued into the apocalyptic tradition, whose main interest was in a systematic study of nature. Using methods from the History of Ideas, Alexander aligns this tradition with forerunners of Greek science in other parts of the Mediterranean shore, with the Ionian philosophers of nature as a prime example. Alexander's thesis is based on the distinction—criticized by some later authors—between the scientific Enoch tradition and the Mosaic tradition, which was less interested in the natural sciences. Alexander initiated the discussion of the Jewish narrative on the history of knowledge by claiming that the myth of the Watchers was designed to disguise the alien origin of sciences like astrology and astronomy by attributing them to a 'Jewish' Enoch. Taken together with Alexander's previous studies on astrology, physiognomy, and magic in the Qumran writings, these studies established a basis for the study of the sciences in Early Judaism, and supplied both the textual and the theoretical infrastructure for the present book.

James VanderKam, a foundational figure in the study of Enochic literature and of its calendars and astronomy in particular, sets out to summarize the scientific teaching of the *Astronomical Book* and analyze its key scientific concepts. The reader is led here along the winding path of Enochic wisdom in its long history of transmission. VanderKam surveys the astronomical teaching of Enoch in the variant textual traditions—Aramaic and Ethiopic—and goes beyond narrower philological concerns to raise two central theoretical questions. He wonders whether the concept of a regular, legalized cosmos as promoted in most of the *Astronomical Book* is compatible with the apocalyptic threat to this order, as demonstrated in the admonition of 1 Enoch 80. This discussion offers a different view of the theme, so central in the present volume, of the encounter between science and its theological infrastructure. After all, for a modern reader it is quite unusual to see science in an apocalyptic framework, and contradictions are certainly due to arise. VanderKam claims that the apocalyptic worldview cannot accommodate this contradiction, and accordingly suggests solving it by assuming multiple authorship. In addition, he critiques Alexander's arguments about the domestication of 'foreign' wisdom in the *Astronomical Book* and his assumption of an opposition between Mosaic and Enochic approaches to knowledge.

The essay by Sanders takes a narrower approach, using a single phrase in the *Astronomical Book* as a window into precisely how "science"—exact knowledge of the physical world—could be imagined as revealed. Sanders notes the passive form of the causative of the standard Aramaic term for 'to see,' חזה, which is used in the *Astronomical Book* to denote the essential formula of revealed knowledge. This linguistic form draws an intriguing continuity between the conceptual world of apocalyptic visionaries and that of the earlier, Priestly, writings from the Pentateuch, where the parallel verbal conjugation denotes revelation of the most central cosmic mystery of the Priestly source: the Tabernacle. At the same time Sanders finds key distinctions between these earlier traditions and the apocalyptic material. The reader thus gains a new lens to view the motivation and the literary genres of ancient Jewish scientific literature.

Stuckenbruck's response is included here because it provides a lucid synthesis and critique of key viewpoints on the earliest Jewish scientific work. Philologically, these include the larger role of verbs of seeing in Aramaic scientific texts and epistemology: was revelation always understood as essentially passive, as Sanders argues, or was there a role for observation, as VanderKam suggests? He also raises the question of the practical role of the *Astonomical Book* in time-keeping and ritual: was it an entirely theoretical text or did it have practical ramifications? Finally, was this cosmic order understood as eternal, or itself a temporary part of a larger historical structure?

The articles by Popović, Ben-Dov and Reed expand our view from the *Astronomical Book* to other manifestations of natural science at Qumran and in early Judaism. They build on the first set of essays in also investigating the stakes of understanding Jewish activity as scientific practice, but expand from a strictly philological view to ask what we can know about the earliest Jewish scientific communities.

Popović attempts to apply to the Jewish material some of the questions of "Social Network Analysis", which has recently gained considerable success in the study of cuneiform scholarship.[21] Criticizing previous approaches to cultural influence that have been limited to the tracing of literary patterns, Sanders wrote, "texts do not create texts; people create them under particular circumstances."[22] Popović takes this agenda further when he points out that "[p]revious research on tracing influences of Babylonian learning in ancient Jewish texts has reflected insufficiently on the specific nature of such cultural encounters and the means of transmission." Thus our task as historians is not complete without asking "how and through whom ancient Jewish scholars got to know about some of the things that Babylonian scholars knew." These two statements call for a larger social and historical examination of what we can say about the actual people who transmitted Babylonian, Egyptian, and Hellenistic knowledge into Jewish hands and created an environment for its reception, an examination for which Popović's critique helps set the agenda.[23]

Popović's is a refreshing update of the study of "provenance" or "origins," which, while indispensable in our field, tends to lead scholars to unproductive attempts at drawing "genealogies of knowledge."[24] Unfortunately there is precious little knowledge about the Jewish literati in contrast to the wealth of cuneiform and Greek material. Popović surveys the range of Jewish scientific material and suggests points of contact with other traditions, yielding a tentative

[21] C. Waerzeggers, "Social Network Analysis of Cuneiform Archives: A New Approach" (forthcoming); E. Robson, "The Production and Dissemination of Scholarly Knowledge" in *The Oxford Handbook of Cuneiform Culture* (ed. K. Radner and E. Robson; Oxford and New-York; Oxford University Press, 2011), 557-576.

[22] Seth Sanders, "The First Tour of Hell: From Neo-Assyrian Propaganda to Early Jewish Revelation" *JANER* 9 (2009): 167.

[23] For a different approach that begins from the range of attested evidence for Babylonian-Aramaic translation from the 9th century through the Hellenistic period, including cuneiform references to tablets of Babylonian scholarly series being copied on a *magallatu* "parchment scroll," see Sanders' forthcoming *Heavenly Journeys and Scholarly Knowledge: The Transformation of Scribal Cultures in Judea and Babylonia* (Brill).

[24] Marc Bloch, "The Idol of Origins" in *The Historian's Craft* (Manchester: Manchester University Press, 1992), 24-28.

map of the contacts that would have been needed to construct this range. His conclusions are rather pessimistic. Little active involvement of Jewish scholars at the forefront of scientific discovery can be posited for the early apocalyptic tradition. Instead Popović speaks of a participation of these Jewish scholars in a shared reservoir of more popular science, including non-mathematical astronomy and other branches of astrology, physiognomy, etc.

The essay by Ben-Dov examines some of the prerequisites for the development of science as they are represented in the early Jewish tradition: the ideals of science (a term borrowed from Amos Funkenstein) as well as the epistemological basis for the production and dissemination of knowledge. Tracing the myths about the birth of knowledge and some statements on its dissemination as they appear in the apocalyptic literature and in the literature of the Yahad, Ben-Dov draws a distinction between these two groups. It seems that the framework of the Yahad (in whatever form it existed) encouraged further development of previously-transmitted knowledge. The Yahad was thus a creative scientific community, marked by some novel scientific productions probably created to answer the needs of the Yahad in the field of diagnostic astrology and physiognomy. The Yahad is thus a good example of a local branch of scientific learning, remote from the centers of learning of the day, which succeeded in creating its own ethos of science, modest as it may have been.

An essay by Annette Yoshiko Reed addresses two of the fundamental concepts underlying the present book. Reed sketches the contemporary scholarly discussion about science as a local, national, product as opposed to the modernistic narrative of universal science. This tension is harnessed in an effort to draw a new cultural history for the emerging discipline of Jewish science in Antiquity. Drawing attention to the lively debate on "the beginnings of Jewish science" in the early Middle Ages,[25] Reed suggests possible forms of continuity between scientific Pseudepigrapha such as *Asaf ha-Rofe*, *Sefer Yetzira* etc. and the earlier material collected in the present volume. Taking into account the generally a-scientific character of rabbinic literature,

[25] Y.T. Langermann, "On the Beginnings of Hebrew Scientific Literature and on Studying History through 'maqbilot' (parallels)", *Aleph* 2 (2002): 169-189.

a continuity between antique and medieval non-rabbinic traditions may prove to be attractive, and will draw attention to other scholarly efforts in this direction, undertaken for example in the study of Mysticism and Hekhalot literature. More broadly, Reed argues that a possible key to understanding ancient science as a coherent enterprise lies in a very broadly attested, perhaps universal feature of it: ancient scientists' *claim* to universality. While the forms ancient sciences took were emphatically local, what may have allowed ancient scientific knowledge to travel so far was that each ancient scientific culture demonstrated an aspiration toward the universal, after an encounter with a universal, cosmopolitan empire.

We suggest two key lessons from the early history of science and Judaism traced here. First, to understand the shifting historical natures of science and religion, we must attend carefully to ancient theories of knowledge. People had different ideas of how to observe and understand the world, and these ideas helped determine what was considered authoritative knowledge. The second is a paradox—but a simple and practical one—about the relationship between the local and the universal.

First, the historical fact is that scientific and mystical claims to knowledge have not always been mutually exclusive. In fact, one may have served the other, as in the revealed science of the *Astronomical Book* of Enoch explored by Alexander, VanderKam and Sanders. In its historical context—the southern Levant of the Hellenistic period—the scientific "content" of the *Astronomical Book* reflected the contemporary common wisdom of popular astronomy and mathematics. Yet Enoch does not observe or calculate to gain this knowledge; rather, he is depicted as learning this scientific knowledge about the universe in precisely the same way that Moses learns the dimensions of the Tabernacle—passively, in a vision that he is *caused* to see. The oppositions we might expect between foreign wisdom versus traditional truth, revelation versus rational knowledge fail to appear—because those oppositions were not theirs.

Second, ancient claims to universal knowledge were made in highly local forms, under highly particular historical circumstances—yet it is precisely the "parochial" qualities of these supposedly universal truths that let us understand them as part of a scientific quest. The simple, and quite practical, paradox is this: each of the scientific cultures under consideration here: Mesopotamian, Judean, and Hellenic, tried to understand and talk about a shared physical universe in a particularly Babylonian, Hebrew, Aramaic, or Greek way. But this very claim to knowledge of a shared universe arose at moments when people were actually becoming part of larger, even "global" networks, networks that inspired and facilitated the tasks of translation and borrowing. The Jewish adaptation of Babylonian astronomy and mathematics explored by Popović happened in Aramaic, the *lingua franca* of the Persian Empire. Similarly, as Reed points out, the Greek and Jewish practice of heurematography, the cross-cultural search for inventors and discoverers, thrived after Alexander's conquest. The ancient sciences we observe here share an essentially local aspiration to the global.

This book marks only the first public scholarly discussion of a new frontier in the history of science and of Judaism in the ancient Near East. But it may already pay three dividends: first, to bring together fundamental data about ancient Jewish science; second, to draw attention to problems in its understanding and to suggest solutions; and finally to broaden a conversation which some of our contributors first began. It is thus no accident that this book is part of ISAW's new open publishing initiative. Whether you find it on the internet or hold a printed copy in your hands anywhere in the world, it is our hope that this book will help inspire new participants in this discussion.

2. Enoch and the Beginnings of Jewish Interest in Natural Science*

Philip S. Alexander

The Jews and Natural Science

In his 1995 monograph *Jewish Thought and Scientific Discovery in Early Modern Europe* David Ruderman discusses the question of Jewish attitudes towards and involvement in science.[1] There is, as has long been noted, an intriguing problem here. Jews in modern times have made a massive contribution to the advancement of the natural sciences, a contribution out of all proportion to their numbers. How is this striking fact to be explained? Are Jews genetically predisposed to be good at science, as some have seriously but implausibly argued. Or does the explanation lie in cultural factors, such as the nature of traditional Jewish education or traditional Jewish love of learning? Or are social forces at work—the desire to escape from exclusion and gain acceptance and influence in the host society which has come to accord great prestige to scientific knowledge? Ruderman problematizes the question by showing that Jewish involvement in science did not begin with emancipation in the nineteenth or early twentieth centuries but can be traced back to the early modern period. In the sixteenth and seventeenth centuries Jews were already studying science, especially medicine, and their numbers were sufficiently large as to have had an impact on Jewish religious thought.[2] As a preface to his study of the early modern period, Ruderman briefly surveys Jewish attitudes towards nature in the Middle Ages but makes no serious attempt to

* I have chosen not to revise this article, though tempted to do so, since it contains the text to which others have reacted. I should, however, make clear that I have now modified some of my views expressed here. I hope to return to the question of early Jewish science in the not too distant future.

[1] David B. Ruderman, *Jewish Thought and Scientific Discovery in Early Modern Europe* (New Haven and London: Yale University Press, 1995).

[2] A case in point is the sixteenth century Italian scholar Ovadiah Sforno whose philosophical and medical training are very evident in his Bible commentaries.

investigate Jewish engagement with science any earlier in Jewish history.

In the present paper I shall attempt, somewhat speculatively, to carry the story further back. I realise that in doing so I run considerable risks. The enterprise may seem grossly anachronistic, and I may project modern debates and modern ideas onto earlier and very different societies. Science today and the society in which it functions look nothing like science or society in the Middle Ages or in antiquity. I am prepared to run that risk. As an historian I am still wedded to the construction of grand narratives. I also hold that analogy is one of the historian's fundamental tools: the past, if it is understandable at all, is accessible only through analogy, through a risky but inevitable process of 'translation' into narratives that make sense in terms of our own experience of the world. And, I would suggest, it is no less meaningful to talk about Jewish science, or the lack of it, in antiquity than it is to talk about Greek or Babylonian science, both of which have been the subject of extensive investigation.[3]

Before we go any further we need a working definition of "science". For our present purposes it is vitally important to avoid one that is too theoretical or exclusive. Since the collapse of Newtonian physics, the supreme exemplar for two centuries of a scientific view of the universe, the nature and definition of science have become philosophically problematic. I am not interested here in this philosophical debate.[4] I am writing as an historian, and as an historian it seems to me obvious that science is a social construct which changes over time. History is littered with sciences—alchemy is a case in point—which have become discredited, and which today are excluded from the scientific curriculum. In the past, however, these subjects were most assuredly regarded as sciences. Though not "true" in the sense that contemporary science is "true", they are 'science-like': they display the assumptions and the articulation of scientific disciplines and in some cases can be shown by historians to have contributed

[3] See, for example, Geoffrey E.R. Lloyd, *Early Greek Science: Thales to Aristotle* (London: Chatto & Windus, 1970).

[4] It is clearly summarized in Menachem Fisch, *Rational Rabbis: Science and Talmudic Culture* (Bloomington and Indianapolis: Indiana University Press, 1997), 1-39.

directly to the rise of modern science. For our present purposes we can identify "science" wherever we find a strong interest in understanding how the physical world works, provided three simple conditions are fulfilled: (1) There is an explicit or implicit assumption that nature is regular and is governed by immutable laws which are accessible to the human mind. (2) An attempt is made to produce a rational model of the physical world which reduces the bewildering complexities of natural phenomena to a small number of underlying primary elements, or to the operation of a small number of fundamental laws. (3) Explicitly or implicitly, a significant element of direct observation of the physical world is involved.

In attempting to trace the earlier history of science two points should be borne in mind. First, experiment plays a major role in modern science. Hypotheses are formulated and experiments devised to test them. In early science, however, experimentation of this type seems to have been rare. It would be a mistake, however, to suppose that if such experimentation is absent, then science is absent. Such an argument has been used in the past to deny that the Greeks possessed any science in any serious sense of that term. Though planned experimentation in the modern sense was comparatively rare, science in antiquity was, to varying degrees, empirically based (one thinks of how the Babylonians' painstaking observations over many centuries of the motions of the heavenly bodies formed the bedrock of early astronomy). And, indeed, I doubt that we can meaningfully talk of science unless there is an element of direct observation of nature. Any proposed scientific model should be, however inadequately or obliquely, either inferred from observation of natural phenomena, or verified by such observation. Second, it may be difficult to distinguish sharply in pre-modern times between science on the one hand and technology and magic on the other. Craftsmen and magicians are, like scientists (at least applied scientists), concerned with exploiting the forces of nature. Technology in the past, as today, has been a great promoter of scientific discovery, but there is surely a distinction to be drawn between the craftsman and the scientist: both may be interested in knowing how things work, but it is the scientist who tries to explain why they work as they do, who formulates theories of nature. The scientist and the magician can be distinguished in a similar way. Magic may seem at times to be predicated on rational assumptions about the mechanistic workings of nature, which the magician can influence by employing the right verbal formula or *materia magica*, but magical texts do not make these assumptions explicit or create a model of nature to make their magical praxis intelligible. The prescriptions of a magician to cure an ailment, even when they contain proposals which are sensible and which may prove efficacious, are an intellectual world away from the rational medicine of a Galen. The scientist, the craftsman and the magician are distinguishable, and were distinguished even in antiquity, though one may merge imperceptibly into the other.[5]

Science and the Talmudic Mind

Jewish involvement in science is not hard to document in the Middle Ages and even in Gaonic times, when Jews engaged seriously with Islamic philosophy.[6] But can we find evidence for scientific interest among Jews in the preceding Talmudic age? Jacob Neusner has argued that not only is science absent from classic Rabbinic Judaism but more fundamentally the logic of Rabbinic discourse, as exemplified in the Mishnah, the foundation document of Rabbinism, is incompatible with scientific modes of thinking and discovery.[7] The

[5] In other words I am using "science" in broadly the sense in which it is used by standard historians of science such as George Sarton (*Introduction to the History of Science*, 5 vols. [Baltimore: Williams and Wilkins, 1927-48]), Lynn Thorndike (*A History of Magic and Experimental Science*, 8 vols, [New York: Macmillan, 1923-58]), and Joseph Needham (*Science and Civiilization in China*, 6 vols, [Cambridge: Cambridge University Press, 1954-96]). If their enterprise is valid, then, *si parva licet componere magnis*, so is mine.

[6] One thinks of Levi ben Gerson with his Jacob's Staff, his modified astrolabe and his criticisms of the Ptolemaic model of planetary motion, or of Abraham ibn Ezra with his interests in mathematics and astrology. For an excellent overview of Jewish science in the Middle Ages see Y. Tzvi Langermann, "Science, Jewish", in *Dictionary of the Middle Ages*, vol. 11 (ed. J.R Strayer; New York: Scribner's, 1989), 89-94. Further, Charles Singer, "Science and Judaism", in *The Jews: Their History Religion and Culture*, vol. III (ed. L. Finkelstein; New York: Schocken, 1960), 216-265, with the postscript by Bernard Goldstein, "The Jewish Contribution to Astronomy in the Middle Ages", 270-275. Further bibliography may be gleaned from Ruderman, *Jewish Thought and Scientific Discovery*, 375-382.

[7] Neusner first published his views in the pamphlet *Why No Science in Judaism?* (New Orleans: Jewish Studies Program of Tulane University, 1987),

implication seems to be that Rabbinical Jews, by their very mental formation, were inhibited from doing science. Menachem Fisch has responded to Neusner's claim by arguing that, on the contrary, the rationality that lies behind the modern scientific enterprise is highly congruent with the rationality of Talmudic discourse.[8] The argument is interesting, but far too essentialist for my purposes.[9] If we descend from this highly abstract, theoretical plane, and look pragmatically at historical realities, we find that there is, in fact, considerable evidence for interest in the workings of nature in Rabbinic literature and Rabbinic society, and, indeed, at a theological level statements occur in the classic Rabbinic sources which can be taken as encouraging and legitimating such an interest. Certain factors may, indeed, have inhibited serious Rabbinic involvement in science. Cosmology (*Ma'aseh Bere'shit*) was famously declared to be an esoteric subject, which could not be expounded before two people (that is to say it could only be studied and taught one-to-one).[10] If this injunction was followed to the

and then in a modified form as "Why No Science in the Mind of Judaism?," in his *The Making of the Mind of Judaism: The Formative Age* (Scholars Press: Atlanta, 1987), 139-160. It is hard to see how the position adopted there squares with his later book *Judaism as Philosophy: The Method and Message of the Mishnah* (Johns Hopkins University Press: Baltimore and London, 1991), in which he argues that "the method and the message of the Mishnah fall into the classification of philosophical methods and messages of the Greco-Roman philosophical tradition. The method is like that of Aristotle; the message, congruent with that of Neo-Platonism" (xi). But if the rationality of the Mishnah is congruent with the rationality of philosophy but not congruent with the rationality of science, it seems to follow that the rationality of philosophy is not congruent with the rationality of science. This is surely a paradoxical conclusion, questionable both in terms of history (which has never sharply differentiated between philosophy and science) and in terms of logic. See further his *Jerusalem and Athens: The Congruity of Talmudic and Classical Philosophy* (Leiden: Brill, 1997). Fisch also notes this problem with Neusner's position (*Rational Rabbis*, 197).

[8] Fisch, *Rational Rabbis*.

[9] I find myself very much in agreement with Ruderman's comment: "Although there are some truth and considerable insight in their [Neusner's and Fisch's] positions, neither offers, to my mind, an adequate historical explanation of the dynamic and complex interactions between science and Judaism. Such theoretical-typological discussions tend to reduce reality to a single categorization or abstract definition, flattening the differences of specific times and places into homogeneous, immutable and predictable entities called science and Judaism." (*Jewish Thought*, 4).

[10] *m. Ḥagigah* 2:1.

letter it would certainly have hampered interest in one fundamental scientific discipline, since it is hard to see how science could have flourished in the rabbinical schools without a free exchange of ideas. There was also the so-called ban on the study of 'Greek wisdom' (*hokhmat yevanit*), which, whatever this term embraced, would surely have included Greek science.[11] Since the dominant science of the Rabbis' day was Greek, to cut oneself off from Greek science was to condemn oneself to a scientific backwater.

However, we do not know to what extent, if at all, these injunctions were observed. And, despite them, knowledge of contemporary science can be found scattered throughout the Talmud and related literature. Given that the Talmud is essentially a book of law, the quantity of this material is actually rather impressive. Take one example, about which I have written at length elsewhere—Rabbinic knowledge of dream interpretation.[12] *Bavli Berakhot* 55a-57b contains within it a substantial dreambook comparable in many ways to the *Oneirocritica* of Artemidorus Daldianus. Dream interpretation was a science in antiquity, with an extensive literature going back to Babylonia. In the case of its most sophisticated practitioners it was empirically based: Artemidorus spent a lifetime travelling around talking to professional dream interpreters and collecting data from them, which they had accumulated from contacts with their clients. Ancient dream interpretation, as Freud saw, anticipated modern psychology in much the same way as alchemy anticipated modern chemistry. The Rabbinic dreambook is not, as one might have expected, based upon the Bible, in which dream messages and dream interpretation play a significant role. It is firmly grounded in contemporary science.

And what are we to make of the *Sefer Yetzirah*? This remarkable and enigmatic little book, which may have been composed in Palestine as early as the third or fourth century CE is essentially a scientific

[11] *m. Sotah* 9:14; *t. ʿAvodah Zarah* 1:20; *b. Menaḥot* 99b; *b. Bava Qamma* 83a. See further my essay "Hellenism and Hellenization as Problematic Historiographical Categories," in *Paul Beyond the Judaism/Hellenism Divide* (ed. Troels Engberg-Pedersen; Louisville, London, Leiden: Westminster John Knox Press, 2001), 63–80.

[12] "Bavli Berakhot 55a-57b: The Talmudic Dreambook in Context," *JJS* 46 (1995): 230-248.

treatise.[13] It was, of course, to become one of the foundation documents, along with the Hekhalot literature, of the Jewish mystical canon, but its mystical interpretation probably does not predate the late twelfth century.[14] Certainly Saadya in his influential commentary on it treats it as a straight-forward treatise on cosmology.[15] *Sefer Yetzirah* proposes essentially an atomic model of the cosmos, in which all the diverse entities of the phenomenal world are seen as different combinations of twenty-two primary elements, symbolized by the twenty-two letters of the Hebrew alphabet. The work is implicitly based on Genesis 1, in that it develops the basic assertion of Genesis that the world was created by the speech of God. God, of course, speaks Hebrew: therefore, it follows that the structure of Hebrew holds the key to the structure of the cosmos. The relationship is symbolic and metaphorical: just as the vast universe of Hebrew discourse is produced by endless combinations of the twenty-two letters of the Hebrew alphabet, so the cosmos can be seen as the product of infinitely varying combinations of twenty-two basic elements. The author of the *Sefer Yetzirah* confirms this model by finding the number twenty-two (in patterns of twelve, seven and three, corresponding to the structure of the Hebrew alphabet) running through the three domains of the cosmos—space, time and the human body. *Sefer Yetzirah* may take Genesis 1 as its starting point but what strikes the reader most forcibly about it is its independence of the Bible. It also illustrates well how thin was the partition between science and magic in antiquity. It is obvious from a close reading that *Sefer Yetzirah* is

[13] See Joseph Dan, "The Three Phases in the History of *Sefer Yetzira*," *FJB* 21 (1994): 7-29. The contrast between *Sefer Yetzirah* and another early Jewish cosmological work, the *Seder Rabba di-Bereshit*, helps to point up just how "scientific" Sefer Yetzirah is. *Seder Rabba di-Bereshit* offers a model of the cosmos, arranged in concentric circles, but it is essentially a symbolic model and apparently arbitrary—symmetry for theological rather than cosmological reasons; it shows no relation to the world as we experience it, or to the science of the day. In no way could *Seder Rabba di-Bereshit* be treated as scientific.

[14] Among the earliest to give it a mystical interpretation was the "Unique Cherub" Circle of the Rhineland: see Joseph Dan, *The "Unique Cherub" Circle* (Mohr Siebeck: Tübingen, 1999), 36-45.

[15] All the major early commentaries on *Sefer Yetzirah* treat it as a scientific work: see Raphael Jospe, "Early Philosophical Commentaries on *Sefer Yetzira*: Some Comments," *REJ* 149 (1990): 369-415.

advocating more than a passive knowledge of the physical world. If one understands how God created the world, then one may be able to control nature and even create new beings. Later Jewish magic seized on this aspect of the Sefer Yetzirah and believed that it held the secret of how to make a homunculus, or *golem*. This may sound like hocus pocus, but it is not too far removed from the modern scientific belief that if one knows nature's laws, then one may be able to control nature and even replicate its processes in the laboratory.

We do not know who wrote *Sefer Yetzirah*, or what his relationship was towards the Rabbinic establishment of his day. There may be an allusion to the work in the Talmud Bavli.[16] Even if there is not, circumstantial evidence suggests that whoever wrote the *Sefer Yetzirah* belonged to what may broadly be termed Rabbinic society. The fact that he wrote in elegant Rabbinic Hebrew points in this direction: it is hard to envisage an audience for such a treatise in third or fourth-century Palestine outside Yeshivah-trained scholars. It should also be borne in mind that Sefer Yetzirah was transmitted to posterity through Rabbinic channels. It seems legitimate, therefore, to take it as evidence for scientific speculation within Rabbinic culture in the Talmudic period.

There is other concrete evidence of scientific interest among Jews in late antiquity. As Raphael Patai has shown Jews were involved in alchemy, possibly from its earliest phases.[17] And they were seriously interested in medicine already in Talmudic times: this is demonstrated by the Talmud itself (Julius Preuss collected and analysed the Talmudic material nearly a hundred years ago[18]), and by *Sefer ha-Refu'ot* attributed to Asaf ha-Rofe, which Elinor Lieber has expended much

[16] *b. Sanhedrin* 67b: "Abaye said: The laws of sorcery are like the laws of Sabbath: certain actions are punished by stoning, certain actions are exempt but forbidden, and certain actions are entirely permitted. He who does a deed is stoned; he who holds the eyes is exempt, yet it is forbidden. What is entirely permitted? Such as the action of Rav Ḥanina and Rav Hoshayah, who spent every Sabbath eve studying the Laws of Creation (*hilkhot yetzirah*), by means of which they created a third-grown calf and ate it."

[17] Raphael Patai, *The Jewish Alchemists* (Princeton: Princeton University Press, 1994).

[18] *Biblisch-talmudische Medizin* (Berlin: S. Karger, 1911; repr. Farnborough, Hants: Gregg International, 1969; English version trans. F. Rosner; New York and London: Sanhedrin Press, 1978).

effort in elucidating for us.[19] Whatever the deep 'logics' of the Talmud may be, it is simply not true as a matter of historical fact to say that Jews—even Rabbinic Jews—were totally uninterested in the natural sciences in late antiquity.

I alluded earlier to the idea that the Rabbinic worldview was not necessarily hostile to the study of nature. This comes out in a striking manner in the pericope which opens Midrash Genesis Rabbah.[20] This pericope is attributed to Rabbi Hoshayah of Caesarea Maritima, who was a contemporary of Origen and may have met and debated with the Christian sage. According to Hoshayah Torah is the blueprint of creation: 'God looked into the Torah and created the world'. Torah is the underlying principle, the *Hokhmah*, of nature. Expressed here is a deep sense that Torah and nature are congruent. But from this it is easy to argue that the study of the one is as legitimate as the study of the other, and, to pick up Fisch's point, that the study of both should be governed by the same rational procedures. Study of nature cannot on this view be inhibited by any fear of a conflict between 'revelation'

[19] See, for example, her excellent survey, "Asaf's Book of Medicines: A Hebrew Encyclopedia of Greek and Jewish Medicine, Possibly Compiled in Byzantium on an Indian Model" in Symposium on Byzantine Medicine, ed. J. Scarborough; *Dumbarton Oaks Papers* 38 (1984): 233-249.

[20] Genesis Rabbah 1:1: "*In the beginning God created* (Gen 1:1). R. Hoshayah commenced [his exposition thus]: *Then I was by Him, as a nursling* (amon); *and I was daily all delight* (Prov 8:30). *Amon* means tutor; *amon* means covered; *amon* means hidden; and some say, *amon* means great. *Amon* means tutor, as you read, *As an* omen *(nursing-father) carries the sucking child* (Num 11:12). *Amon* means covered, as in the verse, Haemunim (*they that were clad*—i.e. covered) *in scarlet* (Lam 4:5). *Amon* means hidden, as in the verse, *And he concealed* (omen) *Hadassah* (Est 2:7). *Amon* means great, as in the verse, *Are you better than No-amon* (Nah 3:8)? which is rendered, Are you better than Alexandria the Great, which is situated among the rivers? Another interpretation: *'amon* means a workman (*uman*). The Torah declares: 'I was the working tool of the Holy One, blessed be he.' In human practice, when a mortal king builds a palace, he builds it not with his own skill but with the skill of an architect. The architect, however, does not build it out of his head, but employs plans and diagrams to know how to arrange the chambers and the wicket doors. Thus God consulted the Torah and created the world, while the Torah declares, *In the beginning God created* (Gen 1:1), *Beginning* here referring to the Torah, as in the verse, *The Lord made me as the beginning of his way* (Prov 8:22)." The *darshan* has correctly identified the basic assertion of Proverbs 8, namely that *Hokhmah* is the underlying, rational order of the universe. He simply assumes that Torah and *Hokhmah* must be identical. See further below.

and 'science', since both are *a priori* based upon the same laws. It is interesting to see how this insight works itself out later in Midrash Rabba. When questions arise about the workings of nature, the Rabbis sometimes find the answers in Torah, and sometimes in direct observation of nature itself.[21] Logically it is all one to them. From a modern scientific point of view this position is naive and untenable. No modern scientist would accept a revealed text as evidence for how nature works: the only valid data for understanding nature are derived from nature itself. But at least this pericope implies that nature functions according to immutable laws that are knowable, and it hints at the idea that the study of nature is as desirable as the study of Torah. The Rabbis' position may not be all that far from that of the devout scientist who believed that when discovering nature's laws he was 'thinking God's thoughts after him.'

Enoch as the Patron of Second Temple Jewish Science

Is it possible to push the story back further still and to find an interest in nature among Jews in the Second Temple period? This brings us to Qumran. In the library of the Qumran sect, both in the sectarian texts and in writings such as the Books of Enoch and Jubilees, which the sect held in high esteem, we find a wealth of interest in the workings of nature and in modelling the cosmos. Different branches of science are represented: cosmology and cosmography, astrology and astronomy (the two disciplines were not clearly distinguished in antiquity), meteorology, calendrical science and physiognomy. The presence of physiognomy comes as something of a surprise, but is should be remembered that physiognomy was a science in antiquity, every bit as much as dream interpretation. It had a well-established technical literature going back to Babylonia, and in various guises (e.g. phrenology) it remained a 'science' down to the nineteenth century.[22] Like dream interpretation, the more sophisticated forms of physiognomy appear to have been based on observation of the

[21] See, for example, Genesis Rabbah 4:4 and 6:8.

[22] See further my essay, "Physiognomy, Initiation and Rank in the Qumran Community," in *Geschichte—Tradition—Reflexion: Festschrift für Martin Hengel zum 70. Geburtstag* (ed. P. Schäfer, H. Cancik and H. Lichtenberger; Tübingen: Mohr Siebeck, 1996), I 385-394.

relationship of character to physical type, and it too anticipated certain aspects of modern science, particularly in the field of psychology.

The calendar in 1 Enoch 72 provides on instructive example of this Second Temple Jewish science.[23] It fits admirably the pragmatic definition of science which I proposed earlier. It clearly attempts to uncover the laws of nature (in this case those governing the sun's motion through the heavens during a solar year), and it provides a rather sophisticated model which integrates in a reasonably satisfactory way a number of precise observations of natural phenomena (see Figure 1). The model is based on the fundamental observation that the sun rises and sets during the year at different points on the eastern and western horizons. From the standpoint of on observer in the northern hemisphere the southernmost point is marked by the winter solstice and the northernmost by the summer solstice. The sun reaches the mid-point between these two extremes at the spring and the autumn equinoxes. The length of daylight and darkness varies according to the position at which the sun rises and sets. The further south the shorter the daylight. The minimum period of light occurs at the winter solstice, the maximum at the summer solstice. The proportions of daylight to darkness are measured on a 18 point scale (implying an eighteen hour day), so that at the equinoxes the ratios are 9:9 and at the solstices 12:6. The eastern and western horizons between the solstice points are divided equally into six matching gates. The sun rises and sets twice in each of these gates as it moves northwards and southwards along the horizons, thus giving twelve months. This division is clearly based not on the movement of the sun but on the movement of the moon, and is dictated by an attempt to fit twelve lunations into the solar year. The author seems to be aware that this cannot be done, so he varies the length of his months. Normal months are thirty days in length, but the months in which the solstices and equinoxes occur are each given an added day. The result is a solar year of 364 days.

[23] See the fundamental discussion by Otto Neugebauer in *The Book of Enoch or I Enoch* (ed. M. Black; Leiden: Brill, 1985), 386-419.

All this may seem rather primitive and obvious, but we should not underestimate how revolutionary both in content and in method such a text must have seemed when it first appeared in Israel. 1 Enoch 72 belongs to the *Book of the Heavenly Luminaries*, the earliest section of 1 Enoch, and is probably to be dated to the late Persian period (around 400 BCE). No earlier text remotely similar to this in its attitude towards nature has survived in the literature of ancient Israel. And we are reasonably sure as to the origins of its doctrine: it is to be found not in earlier Jewish sources but in Babylonian astronomy. It marks the introduction of alien wisdom to Israel.

The Enochic literature contains the highest surviving concentrations of Jewish science from the Second Temple period and much of this science is linked directly or indirectly with the name of the predeluvian patriarch. Enoch is depicted as a great sage, as the fount of all scientific wisdom. Enoch received this wisdom by divine revelation: it was disclosed to him by angels or in visions. There are parallels in Egyptian, Babylonian end other early scientific traditions for presenting science in the form of revelation from the gods, and it is tempting, at first sight, to suppose that in the Books of Enoch this is no more then a literary convention. But there may be more to it than meets the eye. At some point in the evolution of the Enochic literature an author or redactor *must* have known that the knowledge which he was presenting was *not* disclosed by angels but had come from contemporary, non-Jewish sources. Much of it, as we have already noted, appears to have been borrowed from Babylonian science some time in the Persian period. The author or authors, however, did not choose to present their information as simply borrowed from contemporary science. The cosmology of 1 Enoch is in many ways no more crude than the cosmologies of Anaximander or of the other pre-Socratic philosophers, yet *they* do not seem to have resorted to claims of divine revelation. One might equally contrast the attitude of the author of Genesis 1. His account of creation in its present redactional setting within the Pentateuch is implicitly claimed to be divine revelation, but, taken on its own, it makes no such claim. The author tells his tale simply and directly, and does not inform us how he knew such things. Why, then, do the authors of the Enochic literature wrap their doctrine up so comprehensively in the mantle of divine revelations and visions?

I think the answer must, in part, lie in the fact that they were consciously attempting to domesticate within Jewish tradition a body of alien wisdom. They were, at least at the outset, fully aware of the newness of their doctrine—that they were propagating ideas never before heard in Israel. It was for this reason that they insisted so emphatically that in fact they were disclosing old wisdom that was already alluded to in the venerable traditions of Israel. The choice of Enoch as the patron on whom to pin this new teaching is interesting. The brief references to Enoch in Genesis 5 are highly suggestive and hint at a fuller story. It has long been suspected that in the form of the Pentateuch which we now have Enochic material has been edited out. Given that the P-strand of the Pentateuch to which Genesis 5:21-24 belongs is post-exilic, some have suggested that the Enochic literature may actually contain some of these excluded traditions. However, as I understand it, the dynamics of the relationship between the Enochic literature and Genesis 5 is quite different. I have argued elsewhere[24] that the relationship is essentially midrashic. That is to say, the authors of the Enochic literature are exploiting a narrative lacuna in an authoritative text as a way of legitimating new teaching. This implies a certain distance—a discontinuity—between the legitimating text and the doctrine being legitimated. Whatever Genesis 5:21-24 alludes to, it was not Enochic science.

But why precisely Enoch? Why not, for example, Adam? If antiquity was a virtue then surely the progenitor of humankind would have been an obvious choice. Adam is certainly used to validate doctrine in later pseudepigrapha. There is probably an element of opportunism in the choice of Enoch. Midrashists cannot pick and choose: they have to seize on lacunae wherever they happen to occur. But within the grand narrative of Biblical history Enoch suited well the purposes of the Enochic circles. He lay far back in time, before the Flood destroyed human life and disrupted human knowledge.[25] *And* he was older and more venerable than Moses. I have suggested elsewhere

[24] "From Son of Adam to Second God: Transformations of the Biblical Enoch" in *Biblical Figures outside the Bible* (ed. M.E. Stone and T.A. Bergren; Harrisburg, PA: Trinity Press International, 1998), 90-94.

that there is something anti-Mosaic in the Enochic literature.[26] It cannot be accidental that it ignores Moses, and attributes its teaching to someone else. The earliest layers of the Enochic tradition must virtually coincide with the so-called reforms of Ezra. Whatever we may think about the historicity of the Books of Ezra and Nehemiah, they do seem to point to a successful attempt in the Persian period, possibly with Persian royal support, to reconstitute Jewish society in Judah on the basis of the Torah of Moses. That the earliest Enochic writings ignore these developments can hardly be accidental. And there is merit in the suggestion that when the Enochic writings came to be canonised into a Pentateuch, the intent was not simply to imitate the Mosaic Pentateuch, but to challenge it.

Later tradition constantly senses a rivalry between Enoch and Moses. A number of the Enochic traditions were later transferred to Moses in a way that suggests that later writers were uneasy with the powers and authority being granted to Enoch and felt that they should be claimed for Moses. The well-known ambivalence of Rabbinic literature towards Enoch is, I would suggest, motivated by a sense that he is a rival to Moses.[27] There is no way in which one religious system can accommodate two such figures of authority. The circles which stand behind the Books of Enoch were, I would argue, proposing an Enochic paradigm for Judaism in opposition to the emerging Mosaic paradigm—a paradigm based primarily on science as opposed to one based primarily on law. They were innovators: they had taken on board some of the scientific thought of their day and had used it aggressively to promote a new Jewish worldview.

This analysis of the Enochic literature begins to suggest something of the profile of the shadowy group or groups that stand behind these texts. They were, in a sense, modernists. That is to say, they were intellectuals who had access to foreign ideas and were open to the scientific thought of their day. And they were prepared to integrate these novel ideas into a new form of Judaism. But the picture would be incomplete if we failed to spot a deep ambivalence in the

[25] It is also possible that he is being tacitly equated with some Mesopotamian culture-bringer, such as Enmeduranki.
[26] "From Son of Adam to Second God," 117-110.
[27] See especially Genesis Rabbah 25:1.

Enochic literature towards new knowledge. This comes out in a startling contradiction that lies at the heart of the texts as we now have them. There are two great bodies of knowledge referred to in this literature. On the one hand there is the knowledge of nature conveyed by Enoch: this is good. On the other hand there is the body of knowledge conveyed by the Watchers: this is bad. It led directly to the corruption of human society, and to the catastrophe of the Flood. There is no suggestion that the knowledge brought by the Watchers was false knowledge. It was heavenly in origin and mediated by angels, just like Enoch's knowledge. The Watchers were as much culture-bringers as was Enoch. But the knowledge they brought, like the knowledge of Prometheus, was knowledge which Heaven did not want to be disclosed to humankind.

This is intriguing. How can we separate between these two types of knowledge? This brings us to another central theme of the Enochic literature—divine judgement. Side by side with the modernist science of the Enochic literature is a powerful strand of moralizing, conservative ethics. The scientific vision of the cosmos is constantly exploited to ram home the message of divine punishment for sin, and sin is defined not primarily in terms of breaches of the Sinai-covenant (as we would expect in a Moses-orientated worldview), but in terms of life-style, such as the use of cosmetics and jewellery. The vision of the world projected by the Enochic literature is paradoxically *both* modernist *and* reactionary. The circles which produced it saw a strong analogy between the state of society in their own times and the condition of the world before the Flood. There was the same radical corruption. The sins of the Watchers were being repeated in their day, and just as God had responded in the past to such radical evil with overwhelming punishment from which only a righteous remnant escaped, so he was about to do the same again. The world stood once more under the threat of imminent catastrophe. Just as Enoch, the preacher of righteousness, had warned the wicked in his day, so the Enochic circles were warning the wicked in their day and telling them to flee from the wrath to come.

It is noteworthy that the knowledge brought by the Watchers is strongly technological in character: 'magical medicine, incantations, the cutting of roots, and plants ... the making of swords and knives, and shields and breastplates ... bracelets, decorations, shadowing of the eye with antimony, ornamentation, the beautifying of the eyelids, all kinds of precious stones, and all colouring tinctures and alchemy'[28]. For all its modernism 1 Enoch has a whiff of technophobia about it: it is suspicious of technological change. I suspect that this stratum of the literature relates to a period of growing prosperity and materialism, allied to rapid technological development. The situation was not congenial to the conservative mentality of the group. I do not know whether there is anything in the archaeological or the historical record which would enable us to pin-point this time more exactly. I doubt that there is. It is all a matter of subjective perception, which may not correlate all that obviously with historical reality as we can now perceive it. But that the author or authors of these traditions were opposed to social and technological changes taking place in their society is hardly in doubt.

I have already noted that two major images of Enoch dominate the surviving Second Temple period literature—Enoch the Sage who reveals the secrets of nature, and Enoch the Preacher of Righteousness who rebukes the sins of his generation and warns of divine judgement. Corresponding to these two images are the two major themes of 1 Enoch—'science' and 'ethics', descriptions of the cosmos and divine judgement. The two images and the two themes are tightly intertwined in 1 Enoch. Part of the cosmography is devoted to describing the places of punishment of the Watchers and those who follow their evil ways. A close analysis of the literary traditions leaves me in little doubt that the Enoch the Sage and the Culture-bringer is earlier than Enoch the Preacher of Righteousness. Enoch was first exploited in order to validate and domesticate a body of foreign scientific knowledge. Only later—perhaps some one hundred and fifty

[28] 1 Enoch 7:1 + 8:1. I quote here the translation by Ephraim Isaac in *The Old Testament Pseudepigrapha*, vol. I (ed. J.H. Charlesworth; New York, 1983), 16. Isaac's rendering "alchemy" is speculative and based on his Ms A (Kebran 9/II). The Ethiopic literally means "transmutation of the world". It should be noted that the third/fourth century CE alchemical writer Zosimus attributes the introduction of alchemy to the Watchers, and that Enoch came to be closely linked with alchemy through his identification with Hermes Trismegistus. See Patai, *Jewish Alchemists*, 16 and 33.

years later—was this same Enoch the Sage transformed, for reasons which are not entirely clear, into Enoch the Preacher of Righteousness, and the Enochic traditions spun to present a sombre message of impending divine judgement. The same analysis suggests that the Watchers have also undergone a transformation. It is probable that originally they were good—heavenly messengers who descended to earth to bring mankind divine knowledge and to promote the advancement of human culture. When those cultural advances, again for reasons that are no longer apparent, came to be regarded as negative the Watchers were transformed into fallen angels, who had brought forbidden knowledge to mankind and corrupted them, and they were linked with the Sons of God in Genesis 6 who entered into illicit union with the daughters of men.[29]

Science in the Achaemenid Empire

Can we sketch in any more detail the profile of the group or groups that produced the Enochic literature, and relate them more precisely to their times? Most would agree that 1 Enoch has strong links with ancient Jewish wisdom tradition. Within that tradition two contrasting views of physical world can be found in the Persian period.[30] First there is the attitude expressed in the speeches of Yahweh at the end of the Book of Job (chapters 38-41). There Yahweh confronts Job with a catalogue of the wonders of nature. No explanation is offered as to how nature works, only a lyrical description of its mysteries. Indeed the speeches are predicated on the assumption that the ways of God in the physical world are unfathomable to the human mind; the appropriate response to them is one of humility and praise, not study and explanation: 'The Lord answered Job out of the whirlwind, and said, Who is this that darkens counsel without knowledge? Gird up now your loins like a man; for I will demand of you, and declare you unto me. Where were you when I laid the foundations of the earth? Declare if you have understanding.

[29] Jubilees 4:15 hints at this more positive evaluation of the Watchers.

[30] In general see Hartmut Gese, "Wisdom Literature in the Persian Period," in *The Cambridge History of Judaism*, vol. I, *Introduction; The Persian Period* (ed. W.D. Davies and H. L. Finkelstein; Cambridge: Cambridge University Press, 1984), 189-218.

Who determined the measures thereof, if you know? Or who stretched the line upon it? Whereupon were the foundations thereof fastened? Or who laid the cornerstone thereof?' (Job 38:1-6).[31] The author of these lines would surely have regarded it as futile, if not impious, to attempt to discover and to explain how nature works. The dating of the Book of Job is notoriously uncertain, but these speeches probably come from roughly the same time as the Enochic *Book of the Heavenly Luminaries*. The widely accepted fifth-century dating for Job would suit my present argument very well.

The second attitude towards nature is implicit in a text also dating from around the same period—Proverbs 8. There wisdom is personified as the master craftsman who assisted God in the creation of the world. The world is based on wisdom; *hokhmah*, to use the terminology of Heraclitus, a Greek near-contemporary of the author of Proverbs 8, is the Logos of nature[32]—an equation which Philo was later perceptively to develop.[33] But it is this very same wisdom which is said to reside with men, and which they are called upon to embrace and to make their own: 'When God marked out the foundations of the earth,

[31] Qohelet 3:11 is sometimes cited as evidence of scepticism towards man's ability to understand the physical world: את הכול עשה יפה בעתו גם את העלם נתן בלבם מבלי אשר לא ימצא האדם את המעשה אשר עשה האלהים מראש ועד סוף, but the text is a well-known crux. The NRSV probably correctly conveys the sense: 'He has made everything suitable for its time; moreover he has put a sense of past and future into their minds, yet they cannot find out what God has done from the beginning to the end.'

[32] See, e.g., Diels-Kranz[11] Frgs 1, 2, 50 and 114, with the commentary in *Heraclitus: The Cosmic Fragments* (ed. G.S. Kirk; Cambridge: Cambridge University Press, 1962), 32-71: Frg. 1: "Of the Logos which is as I describe it men always prove to be uncomprehending, both before they have heard it and once they have heard it. For although all things happen according to this Logos, they [men] are like people of no experience, even when they experience such words and deeds as I explain, when I distinguish each thing according to its constitution and declare how it is; but the rest of men fail to notice what they do after they wake up just as they forget what they do when asleep." Frg. 2: "But although the Logos is common the many live as though they had a private understanding.' Frg. 50: 'Listening not to me but to the Logos it is wise to agree [*homologein*] that all things are one." Frg. 114: "Those who speak with sense must rely on what is common to all, as a city must rely on its law, and with much greater reliance: for all laws of men are nourished by one law, the divine law; for it has as much power as it wishes and is sufficient for all and is still left over" (translations by Kirk).

[33] Philo's Logos is indebted not only to the Platonic Logos but to the ancient Hebrew concept of Wisdom.

then I [Wisdom] was by him, as a master craftsman (אמון); and I was daily his delight, rejoicing always before him; rejoicing in his habitable earth; and my delight was with the sons of men. Now, therefore, my sons, hearken unto me: for blessed are those who keep my ways. Hear instruction, and be wise' (Prov 8:29-33). It would not be hard to deduce from this passage, though the text does not explicitly do so, that it is perfectly possible, legitimate, and, indeed, desirable to study the wisdom that fashioned the world.

The circles that inaugurated the Enoch tradition took the Proverbs 8 line. They were as impressed as the author of Job by the wonders of nature, but they saw this as no bar to studying or to explaining how nature worked. They had, as Isaac Newton would have appreciated, the attitude of the true scientist: awe before nature, but at the same time an irresistible urge to probe its mysteries, and when the mystery is explained, the awe is not dispelled but only deepened. The circles that stand behind 1 Enoch seem to have emerged in Israel in the later Persian period. Their science, as we have already noted, appears to have been drawn largely from Babylonian sources. This is hardly surprising. Babylonia dominated early science, particularly the exact sciences,[34] and Babylonian scientific ideas were certainly transmitted westwards to the Greeks, and, as 1 Enoch and related texts make clear, to the Jews.

This westward transmission of Babylonian ideas would have been facilitated by the political and cultural conditions that prevailed under the Persian Empire. It is surely highly significant that the language of the Enochic traditions is Aramaic. This fact is usually not paid the attention it deserves. In the fifth or early fourth century BCE in Judah it was probably something of an innovation to write a work such as the *Book of the Heavenly Luminaries* in Aramaic. Aramaic was, indeed, spoken by many in post-exilic Judah (though not precisely the Aramaic of 1 Enoch, which is in a high, literary register), but Hebrew was by no means dead, and it remained unquestionably the language of literature. The reason for the Aramaic is quite simple: Aramaic was the *lingua franca* of the Persian Empire for administrative and diplomatic

[34] See Otto Neugebauer, *The Exact Sciences in Antiquity* (New York: Dover, 1969, 2nd ed).

purposes; and it probably functioned as the language of international culture as well. It was in Aramaic that the Enochic circles received the Babylonian scientific traditions; it was in Aramaic that they preserved them. The ideas were new in Israel and Hebrew as yet lacked a technical, scientific vocabulary in which to express them. An analogous situation arose in the early Middle Ages, when Jews began to write in Arabic, not so much because it was the vernacular, but because it was the language of high culture and science, and Hebrew had yet to develop a scientific vocabulary. But the Arabic scientific literature which the Jews read, was not, at least initially, transmitting *Arabic* ideas, but rather *Greek* ideas in Arabic dress.

The Enochic circles were obviously well educated: they had mastered literary Aramaic and they had access to foreign literature. Most likely, therefore, they belonged to the scribal and priestly classes in Jerusalem. They seem to have retained some sort of existence over a considerable period of time, and to have continued to work on and develop the Enochic traditions. That development, as we have noted, earlier was increasingly in a moralising direction. The science was put directly to the service of religion, to support a message of impending divine judgement. The Enochic literature was, as we know, taken up by the Qumran community, for reasons which are not immediately apparent. The Qumran community had its own sectarian view of the world, focused sharply on a model of time which portrayed nature as moving purposefully towards a climactic final conflict between the cosmic principles of light and darkness (a notion very probably indebted to Persian thought). The message of impending catastrophic judgement in the Enochic literature was doubtless congenial to them in a general way, but there is little sign of the detailed Enochic cosmographies playing a central role in sectarian thinking.

The standard explanation of Qumranian interest in Enoch is that the Qumranians, in opposition to the Jerusalem priesthood, had adopted the Enochic solar calendar, and needed both the Enochic science and the authority of the Enochic literature to sustain its position. However, this view is not without problems. It is likely that the Enochic 364-day solar calendar did, originally, represent an attempt to reform the Jewish calendar, in accordance with the best science of the day. It is possible that the new calendar was presented as a way of living more in accord with the laws of nature and of God. But, of course, the calendar does not work, and it would not have taken long for people using it to notice that it does not work: without correction it should have been obvious within thirty years that it was badly out of synch with nature. And in a community that may have lasted almost two hundred years, the discrepancy would have become glaring and disastrous. The calendar may have been retained as an ideal model of time—a kind of model not unknown to modern science. It may have come to represent how time ideally should run, and perhaps would run in the future, when the natural order was no longer disturbed by evil. It is, of course, possible that as a community of scholars, the Qumranians valued the Enochic texts for their own sake as learned, and, indeed, edifying literature, without being too deeply influenced by them. But the simplest explanation is surely that Enoch features at Qumran because the circles who founded Qumran were linked in some way to the circles that studied the Enochic tradition. Enoch was part of their intellectual baggage. The Jerusalem Temple in the Second Temple period was probably a locus not just of ritual, but of a vigorous intellectual life, and may have housed a school or schools. This should, in principle, cause no surprise: great temples had from hoary antiquity been centres of learning in the Near East. Qumran was founded by renegade Jerusalem priests. The founders of Qumran were associated with the school, or the circle, in the Jerusalem Temple which had preserved and studied the Enochic literature, and they brought copies of the texts with them from there to Qumran.

Be this as it may, if my analysis is even half correct, then it points to a rather interesting conclusion. Sometime in the late Persian period, say around 450-400 BCE, under the influence of Persian and, ultimately, of Babylonian ideas, Jews for the first time became interested in producing scientific models of the workings of the natural world. Though to some extent anticipated by the simplified, largely demythologized account of the origin of the world in Genesis 1 and by the assertion that behind the natural order lies a *hokhmah* accessible to the human mind in Proverbs 8, the approach to nature displayed in the Enochic *Book of the Heavenly Luminaries* is unprecedented in Jewish literature. It seems to mark a turning-point in Jewish intellectual history—the emergence, for the first time, of what might properly be called a scientific attitude.

One might compare the analogous intellectual revolution which had taken place about a hundred years earlier in the Greek world, under the influence, possibly, of the similar intellectual stimuli. I refer to the rise of the Ionian school of Greek philosophy and science. The Ionians too produced new, rational models of the cosmos—models little more sophisticated than those of the Jewish Enochic circles, but which in the *Heilsgeschichte* of western civilization are traditionally seen as the beginnings of Greek, and indeed of European, science. In both cases—the Jewish and the Greek—the new models of the universe marked a qualitative break from pre-existing mythical and epic pictures of the world. In the case of the Greeks those earlier pictures were to be found in Homer and in Hesiod; in the case of the Jews they were enshrined primarily in the opening chapters of Genesis. In both cases some reference was made in the new models to the old mythical ideas. This is certainly the case in 1 Enoch which, at least in its present form, bears a loosely exegetical relationship to Genesis 1, but there seem to have been allusions to the traditional cosmogonies in the Ionian cosmologies as well.

The question of eastern influences on the Ionians is controversial, but in the wake of the "orientalizing revolution"[35] it is now widely acknowledged that external ideas played a significant role in the development of Ionian thought.[36] The source or sources of those ideas

[35] See Walter Burkert, *The Orientalizing Revolution: Near Eastern Influences on Greek Culture in the Early Archaic Age*, (trans. M. E. Pinder and W. Burkert; Cambridge, MA: Harvard University Press, 1992).

[36] For an overview of the question see Edward Hussey, *The Presocratics* (London: Duckworth, 1972), 1-31. Though his account is generally balanced, Hussey still wants to reserve something unique for the Ionians. He maintains that 'the core of the Milesian revolution, namely the development of a reformed theology based on general principles, and the correlative vision of a universe governed by universal law, cannot be paralleled, as yet, from anywhere outside Ionia' (p. 29). But if the argument of the present paper is correct then a group of Jews seem to have reached more or less the same position at more or less the same time as the Ionians. Hussey notes the attitude towards nature in Job and shrewdly compares it with Pindar, but he misses the significance of Proverbs 8, or even of the heavily demythologized account of the origins of the world in Genesis 1,

is not entirely clear. The Greeks themselves looked to Egypt, but modern scholarship points more emphatically to Babylonia and Persia. The Ionians are unlikely to have known much about Persian ideas before 540 when the Persians reached the Aegean coast. Only in the time of Heraclitus do we find more or less convincing evidence of distinctively Iranian influences on Greek thought.[37] However, the rather sharp distinctions often draw between Babylonian, Persian, Egyptian and Canaanite thought may be misleading. The picture that is now emerging is of an increasingly internationalized culture in the Near East in the sixth and fifth centuries BCE, with a remarkably free interchange of ideas among the educated elites. This may have come about in part through migration of individuals (Martin West assigns a major role to wandering Magi[38]), but politics, and concomitant

which, as we noted earlier, in itself makes no claim to prophetic revelation. Hussey's grasp of ancient Jewish sources is notably uninformed. He also tends to tie the Milesian revolution too tightly to the political conditions of the city-state. This seems to imply that science and philosophy can only flourish under "democracy". The Milesian philosophers were almost certainly from a rather different social background from that of the members of our Enochic circles in Jerusalem. The former seem to have been men of affairs, with no obvious religious role, whereas the latter were probably priests. But it would be wrong to deduce from this that these two groups would automatically have held fundamentally different views of the world, and that the priests could not have been rational or scientific. Nevertheless Hussey's willingness seriously to entertain the possibility of eastern influences on the Ionians marks a seismic shift from the older histories of the Presocratics such as Geoffrey S. Kirk and John E. Raven, *The Presocratic Philosophers* (Cambridge: Cambridge University Press, 1957, 1st ed.) and William K.C. Guthrie, *A History of Greek Philosophy*, vols I-II (Cambridge: Cambridge University Press, 1962-65).

[37] See Martin L. West, *Early Greek Philosophy and the Orient* (Clarendon Press: Oxford, 1971), 111-202.

[38] *Early Greek Philosophy and the Orient*. chap. 7, "The Gift of the Magi". There are persistent rumours in the Greek doxographical tradition that some of the Ionian thinkers were actually themselves of 'oriental' stock, but it hard to know what credence to give to these traditions. West provides a wealth of oriental parallels to the Presocratics, some more convincing than others, which build cumulatively into a conclusive case. However, his historical explanation of these parallels leaves something to be desired. He is a Pan-Iranist, who paints a rather romantic picture of Magi scattering from Persia eastwards into India, where they lay the foundations of Indian philosophy and westwards into Asia Minor where they profoundly shape Greek thought. And like most writers on these subjects he ignores the linguistic question: through the medium of which language did these ideas spread?

linguistic and commercial factors, were probably more decisive. In the sixth century the Babylonians dominated the Near East politically: that doubtless fostered trade and gave the whole region a *lingua franca*, Babylonian. When the Persians succeeded the Babylonians as the political masters, Aramaic replaced Babylonian as the *lingua franca*. This almost certainly did not mean the end of Babylonian cultural influence, since Babylonian ideas were probably carried over into Aramaic. Local intellectual elites were able to buy into this international culture by learning Babylonian and Aramaic. Greeks would have had a flying start in the case of Aramaic, given that it was written in basically the same alphabet as they had adapted for their own language.

What I am suggesting, then, is that we can identify at least two groups within this international culture, one in Miletus in western Anatolia and one in Jerusalem in Judah, which independently of each other but influenced by the free circulation of ideas through the Levant and the Near East, developed a view of nature which within their own societies was radically new and which can for the first time be meaningfully labelled as 'scientific'. This interest in nature, inaugurated among Jews in the Persian period, continued in fits and starts down to the Middle Ages. In the Middle Ages and early modern times, as Ruderman has shown, it gathered pace. In the nineteenth century, as a result of political emancipation, many Jews again rediscovered the natural world. The result, in the twentieth century has been some of the greatest achievements of scientific thought. From Enoch to Einstein is a long and tortuous road. At times the traces are scuffed and the track almost disappears. But it looks like a road which the historian of Judaism could and should map along the whole of its length.[39]

[39] Whether or not it is meaningful to talk about *Jewish* science cannot be discussed here. For the historian of Judaism the important point is the extent to which religious ideas and scientific ideas interacted in Judaism. See further Patai, *Jewish Alchemists*, 517-518.

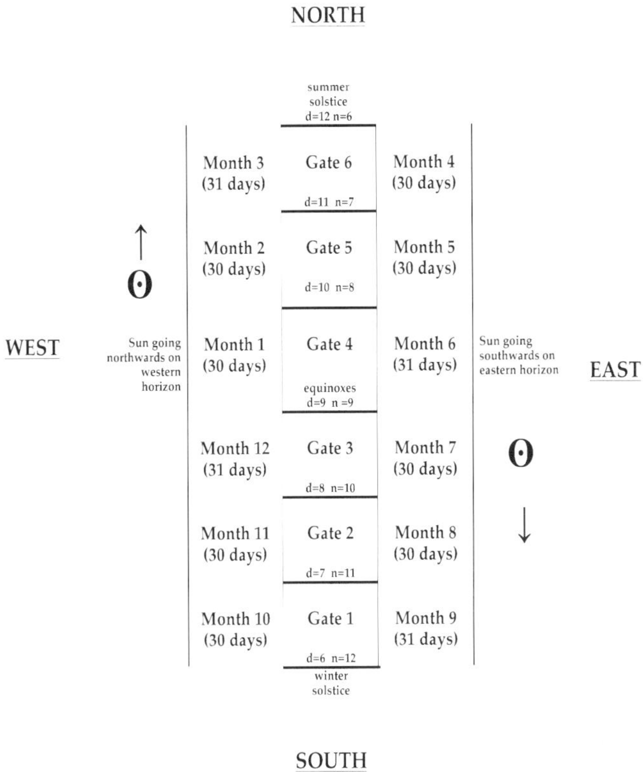

d = day n = night

Months 1-3 = FIRST SEASON Months 4-6 = SECOND SEASON

Months 7-9 = THIRD SEASON Months 10-12 = FOURTH SEASON

Figure 1: The Enochic Model of the Sun's Motion

3. Enoch's Science

James VanderKam

Enoch, so far as we know, was the first hero in the Jewish tradition with whom scientific material was associated. His area of scientific research and writing was astronomy, and an entire booklet containing his teachings on the subject has been preserved. The *Astronomical Book* of Enoch or the *Book of the Luminaries* survives on a series of fragments from four manuscripts found in Qumran cave 4 (4Q208-211)[1] and to a greater extent though in a different form in a large number of Ethiopic copies. The Aramaic fragments preserve text in the original language of the composition; the Ethiopic version is a translation of a Greek rendering of the Aramaic. Virtually nothing of that intermediate Greek version is extant so it will play only a modest part in this essay.[2]

Enoch's scientific concerns—or, as they are presented in the texts, the revelation to him of scientific data—come to expression in other places than the *Astronomical Book*. For example, in the *Book of Watchers* (1 Enoch 1-36) he mentions teachings about some astronomical subjects (among others) by angels who sinned in the way they made the information available to people (1 Enoch 8). In the same booklet

[1] Joseph T. Milik made available much of the evidence in preliminary form in his *The Books of Enoch: Aramaic Fragments of Qumrân Cave 4* (Oxford: Clarendon Press, 1976), 273-297 with pls. XXV-XXX. Publication of 4Q208-209 was completed by Eibert J.C. Tigchelaar and F. García Martínez, "208-209. 4QAstronomical Enoch[a-b] ar," in *Qumran Cave 4 XXVI: Miscellanea, Part 1* (DJD 36; J. VanderKam and M. Brady, consulting editors; Oxford: Clarendon, 2000), 95-171. Henryk Drawnel has now produced a thorough edition of the four Aramaic mss.: *The Aramaic Astronomical Book (4Q208-4Q211) From Qumran: Text, Translation, and Commentary* (Oxford: Oxford University Press, 2011).

[2] Milik identified some Greek fragments as containing text from the *Astronomical Book* of Enoch (Joseph T. Milik, "Fragments grecs du livre d'Hénoch [P. Oxy. XVII 2069]," *Chronique d'Égypte* 46 [1971]: 321-348, especially 333-341); Randall Chesnutt has more recently examined the fragments and strengthened the case for identifying them as from the Enochic book: "Oxyrhynchus Papyrus 2069 and the Compositional History of 1 Enoch," *JBL* 129 (2010): 485-505.

Enoch himself tours the cosmos and views its structures, but the overlapping sections (especially chs. 33-36) offer little that adds to the store of his understanding of the way the universe works. In the *Book of Parables* (1 Enoch 37-71) he again sees or names parts of the universe (41:3-8; 43-44; 59:1-3; 60:11-22; 69:13-25; 71:4) as he travels with angels, but in none of these passages, although they treat some of the same topics as the *Astronomical Book*, is there the sustained attention to the paths of the luminaries and the measure of their movements that one finds in the astronomical chapters.[3]

Features of the versions in which the *Astronomical Book* has come down to us confront readers with a challenge in employing it for close study. The Aramaic copies are badly damaged so that only pieces have survived, usually small ones and ones often difficult to read. From the remains it appears that a systematic, list-like presentation of lunar data was at the heart of the composition. For date after date the texts record the time during which the moon was visible or invisible and the amount of the lunar surface that was illuminated or not illuminated. The Ethiopic manuscripts appear to preserve a complete composition, but the relation between the Ge'ez text and the Aramaic is decidedly problematic. So, for example, the Ethiopic translation includes only an abbreviated version of the lunar material that appears to be so ample in the Aramaic text. Nonetheless, the two share a number of sections and traits so that one can draw some conclusions from the work.[4]

The goal of this paper is to ask some basic questions about the nature of the science one finds in the astronomical work associated with Enoch: the data in it, the ways in which they are presented, and their sources. Once that material is before us, there will be consideration of broader issues in connection with ancient Jewish science, including the *Astronomical Book* of Enoch.

[3] For comparisons of these sections with the *Astronomical Book*, see my survey in George Nickelsburg and James C. VanderKam, *1 Enoch 2: A Commentary on the Book of 1 Enoch, Chapters 37-82* (Hermeneia; Minneapolis: Fortress, 2011), 390-398.

[4] See my analysis of the relation between the Aramaic and Ethiopic versions in *1 Enoch 2*, 351-357.

I. The Book Itself

First let me furnish a sketch of what is contained in this earliest Jewish scientific work. Because of the differences between the two versions—the Aramaic and the Ethiopic—the parts they share will be the focus. Whatever its earliest shape may have been, at some point the work came to encompass several different elements, all encased in the literary frame of a father instructing his son, possibly in a testamentary situation (76:14; cf. 4Q209 23 1-2 and 4Q210 ii 14).

A. Lunar data: Most of the extant text on the Aramaic fragments offers sections detailing the amount of light and the time of visibility the moon has each night throughout successive months. All the letters and words surviving from 4Q208 and everything on 4Q209 frgs. 1-22, 30-40 (?) fit into this category; there is also lunar material in other fragments, but it is not part of these particular sections (e.g., 4Q210 iii). It is as if an author has taken tabular data and written it in prose form. The lunar data are attested in highly truncated form in the Ethiopic version in 73:1–74:9; 78:6-17. In both versions the moon passes through a series of *gates* on the horizon; the gates are, as Otto Neugebauer showed,[5] equal segments or arcs of the horizon. The gates are mentioned frequently in the Aramaic fragments, while the Ethiopic text says there are six on the eastern and six on the western horizon (e.g., 72:3). The highest ordinal with the noun *gate* in the Aramaic is *sixth* (see 4Q208 33 2), suggesting that the same structure underlies its data. In both versions the amount of the moon's surface illuminated and darkened and the time of the moon's visibility and invisibility are expressed in fourteenths: the Aramaic fragments speak of half seventh parts (cf. 4Q210 iii 6), and the Ethiopic uses fourteenth parts.[6] The lunar data are coordinated with dates in the months. H. Drawnel has analyzed the lunar sections of the Aramaic fragments in great detail and concluded that the tables contain these details: for the waxing phase of the moon, they note: A. the time involved is at night, B. the time from sunset to moonset, C. the setting of the moon, D. the time from moonset to sunrise, E. from sunrise to moonrise, F. an equation

[5] "Notes on Ethiopic Astronomy," *Or* 33 (1964): 49-71, especially 51-58.

[6] There is too little left of the Greek version to be sure about the point, but it may preserve the same division into fourteen parts as in the Ethiopic version (see Chesnutt, "Oxyrhynchus Papyrus 2069," 493-494).

(regarding the amount of light), and G. from moonrise to sunset. There is another pattern when the waning phase of the moon is under consideration.[7]

The fact that both versions operate with fourteenths means that questions arise about the length of a lunation: does the system presuppose that a month lasts 28 days? There are a couple of places where treatments of the middle and the end of the month are almost preserved, but, typically, just where one would like a few more words, the fragments break off. For instance, in 4Q209 6 7-9 the writer describes the 28th of a month (the number is preserved entirely) and mentions that a half of a seventh part of the lunar surface is illuminated. During that night the remaining half of a seventh is obscured so that the moon is devoid of light, hidden with the sun. This is the last preserved line on the fragment (apart from a couple of letters in line 10) so that the treatment of the next day is lost. 4Q209 8 3 mentions night fifteen and apparently indicates that on the preceding night the moon was visible the full time; the next surviving letters (line 4) may refer to the sun and its course. See also 4Q210 iii 4 which speaks of the fourteenth day, while line 5 lists the fifteenth day and says its light is complete.[8]

B. Solar data: The annual course of the sun, month by month, is the subject of the opening chapter in the Ethiopic version (ch. 72). There the sun, like the moon and stars, moves through the six gates on the eastern and western horizons during a year of twelve months. Each month consists of 30 days except numbers 3, 6, 9, and 12 that have a thirty-first day for a total of 364 days. Several times in the book

[7] Henryk Drawnel, "Moon Computation in the Aramaic Astronomical Book," *RQ* 23/89 (2007): 3-41; see pp. 35-36 for the two tables. For a slightly revised version of the tables, see *The Aramaic Astronomical Book*, 243-259.

[8] Drawnel (*The Aramaic Astronomical Book*, 285-290) discerns two patterns, one for months with the full moon on day 14 and one for months with it on day 15 (see also pp. 421-424). In 1 Enoch 74:10-16 there are 354 days in twelve lunations; 78:15-16 refers to six months of 30 days and six of 29.

the writer mentions that the year lasts 364 days and chides those who think it consists of 360 days only (see ch. 75; 82:4-8).[9] The sun is less prominent in the surviving parts of the Aramaic version, but the noun occurs nine times. In one case, only the first letter of the word survives; for three or four of the remaining eight passages so little of the context is extant that no meaning can be gleaned from them. As for the better preserved sections one can tell that the writer spoke of the sun moving through various *sections* (4Q209 7 iii 1-2, 5) and that it goes back over the same course through which it had come (4Q209 7 iii 5). In addition, the text must have compared the number of days in a certain period measured by the sun with one measured by the moon because it says the moon has a lack or deficit in comparison with the sun (4Q209 26 3).[10] It also deals with the relative movements of the sun and moon, as it mentions that the moon completely lacks light on its surface when it sets with the sun (4Q209 6 9). None of the Aramaic fragments evidences a text such as 1 Enoch 72 which is almost totally devoted to the annual path of the sun through the gates on the horizon.[11]

C. Geographical data: Both versions contain a section regarding the twelve gates for the twelve winds, three in each of the four cardinal directions (1 Enoch 76; 4Q209 23 1-2; 4Q210 ii 1-10, 14; the number *twelve* for the gates is preserved on 4Q210 ii 14, as is the number *four* for the quarters or directions), and a unit about the four quarters of the earth and its seven great mountains (1 Enoch 77; 4Q209

[9] There is ample Ancient Near Eastern evidence for a schematic year of 360 days, and the Enochic astronomy seems also to presuppose the same number as the gate system implies, although the author argues the year really does consist of 364 days. Whether the situation is to be explained as evidence for a redaction of an earlier form of Enochic astronomy (see the survey in J. Ben-Dov, *Head of All Years: Astronomy and Calendars at Qumran in their Ancient Context* [STDJ 78; Leiden: Brill, 2008], 32-37) or as an incomplete revision of the earlier system that was never part of the *Astronomical Book* would be difficult to determine. However one views the development, there is no denying that the gate system in relation to the annual movements of the sun fits a 360-day year better than the one of 364 days.

[10] 1 Enoch 74:10-16 compares the lengths of the solar and lunar years, but no Aramaic or Greek fragment corresponding with this section has survived.

[11] Milik (*The Books of Enoch*, 273) thought the Aramaic form of the astronomical work may have included a "broad introduction (approximately equivalent to En. 72)", but there is no trace of such a section in the surviving fragments.

23 3-10; 4Q210 ii [14]-20). These are sections in which there is a larger measure of overlap between the versions than in the lunar material, although there are many differences as well. Geographical sections in astronomical/astrological works are expected, as the sun, moon, and stars were thought to affect events in different places on the earth.[12] Enoch's astronomical work was influenced by such models but modifies them in a non-divinatory direction.[13] The fact that in the booklet he receives revelations about astronomical topics from an angel would not have appeared strange in the ancient context where Enoch's astronomical writing first appeared.

D. Patterns: In all of these sections there are certain patterns that can be summarized as follows.

1. The laws governing the creation are unchanging.[14]

a. 1 Enoch 72:1, a passage not reflected in the Aramaic fragments but entirely consistent with them, enunciates the point clearly: "The book about the motion of the heavenly luminaries, all as they are in their kinds, their jurisdiction, their time, their name, their origins, and their months which Uriel, the holy angel who was with me (and) who is their leader, showed me. The entire book about them, as it is, he showed me and how every year of the world will be forever, until a new creation lasting forever is made."[15] The principle of a stable world

[12] See James VanderKam, *Enoch and the Growth of an Apocalyptic Tradition* (CBQMS 16; Washington, DC: Catholic Biblical Association of America, 1984), 52-75, 89-104.

[13] The relation between ancient astrology and what looks more like astronomy to us is a complicated one. As Francesca Rochberg argues, even scribes of the highly technical astronomical works in the last centuries of cuneiform literature were committed to "tradition and the idea of the divine nature of knowledge" (*The Heavenly Writing: Divination, Horoscopy, and Astronomy in Mesopotamian Culture* [Cambridge: Cambridge University Press, 2004], 298; see her entire discussion on pp. 237-299). The same data were shared by the omen series and the more mathematical astronomical works, though they were applied to distinguishable ends. See also her comment (p. 96): "In a preliminary way, however, it might be suggested that, apart from the divinatory purpose of the omen series, the status of these series as systematically acquired corpora of 'what was known' justifies an identification as science."

[14] A similar view comes to expression in other Enochic works (see, for example, 1 Enoch 2:1–5:3; 69:13-25).

[15] Translations of 1 Enoch are from Geroge W.E. Nickelsburg and James C. VanderKam, *1 Enoch: The Hermeneia Translation* (Minneapolis: Fortress, 2012).

is one reason why most of ch. 80, which predicts the dissolution of the created order, is unlikely to be an original part of the booklet.

b. The phenomena described in the booklet do not deviate from the course or pattern. For instance, the lunar data in the Aramaic fragments appear to be set, fully predictable lists. The numbers move by one-fourteenths (halves of a seventh) between zero and one; they never deviate. The same could be said for the solar data in the Ethiopic version and for the lunar material although it is only partially preserved and has a few difficult passages.

2. In line with its ideal, schematic character, the book frequently uses a small set of numbers: 3, 4, 7, 12, and 14:

a. 3: Each season lasts three months (82:11), and each of the four cardinal directions has three gates through which three winds blow (76:1-3; cf. 4Q210 ii 1-10).

b. 4: There are four cardinal directions and four parts of the earth (77:1-3 4Q209 23 3-9; 4Q210 ii [14]-19); there are also four seasons (and four additional days in the solar year in the Ethiopic version)

c. 7: Though the week is not an important unit in either version, the solar year lasts exactly 52 of them; there are seven great mountains, rivers, and islands in the earth (77:4-8; see 4Q209 23 10; 4Q210 ii 20); the light of the sun is seven times that of the moon (72:36; 73:3; 78:4); and the Aramaic version speaks repeatedly of sevenths when dealing with the moon. Of course, Enoch himself was the seventh from Adam.

d. 12: There are 12 months, 12 gates, six on each horizon, through which the sun, moon, and stars pass in their annual cycles (ch. 72); there are also 12 openings in the sun's disc (75:4), 12 gates for the 12 kinds of winds, three in each of the cardinal directions (ch. 76; cf. 4Q210 ii 1-10).

e. 14: There are 14 units of the moon's surface that can be illuminated, and there are 14 units of time the moon is visible/invisible (e.g. 74:1-9; 78:6-17; see 4Q210 iii 3-9). Each of these corresponds with one date in the waxing and waning phases of the moon. (The solar day has 18 parts, but with 6 and 12 being the extremes [ch. 72].)

3. Observation: A repeated phenomenon in the more completely preserved Ethiopic version is Enoch's observation of the data he describes. The same can be glimpsed for the Aramaic as well though to a lesser extent, perhaps due to its poor state of preservation. The angel Uriel reveals aspects of nature to Enoch, but the patriarch says repeatedly that he saw them—they were not simply dictated to him (see below for the references). There is no explicit mention of his actually being on a journey with Uriel in the astronomical work (in ch. 81 he is with seven angels but the chapter may belong to an editorial layer), though he may have been; otherwise he must have been located in a very special place to be able to see what he claims to have seen.

The statements about observing or seeing the features concerning which he reports should be subdivided into two principal categories

a. Enoch says that Uriel showed phenomena to him

a book about the motions of the luminaries (72:1)

the moon as it carries out its prescribed course, its positions and light (74:2)

sign (= an extra day each season), seasons, year, days (75:3)

12 gates open on the disc of the sun's chariot (75:4)

a law regarding light from the sun that illumines the moon (78:10)

all the laws, etc. of the stars (79:2)

the luminaries (79:6)

the luminaries, months, festivals, years, days (82:7) (cf. 80:1; 81:1)

b. Enoch says he saw certain objects

six gates (72:3)

the law about the moon (73:1)

another course and law (74:1: this example may clarify what is meant by seeing a law: "Another course and law I saw for it; by that law it carries out its monthly course." Enoch sees the motions of the relevant heavenly body and perceives its law or has it revealed to him by observing that motion)[16]

[16] It seems unlikely that Uriel is showing Enoch information inscribed on the heavenly tablets. In 1 Enoch 72-82, those tablets are mention only in 81:1-2 where they have nothing to do with astronomical information. There are also strong reasons for thinking that ch. 81 is an editorial unit and not an original part of the composition (see VanderKam in Nickelsburg and VanderKam, *1 Enoch 2*, 531-536).

the relative positions of the sun and moon (74:9)
chariots in the sky (75:8)
twelve gates open for the winds (76:1)
seven great mountains (77:4)
seven great rivers (77:5)
seven great islands (77:8)
everything that precedes this point in the text (80:1) (cf. 81:2 where he looks at the heavenly tablets)

c. For the sake of completeness, it should be noted that Enoch in turn shows the information to his son Methuselah

I have shown you everything (all the material in 72-76? 76:14); it is also in the Aramaic (4Q209 23 1-2; 4Q210 ii 14)

I have shown you everything (79:1); this too is reflected in the Aramaic (see 4Q209 26 6-7)

I am telling you these things (82:1) (cf. 82:2; 81:5)

The Aramaic copies preserve a few such references and certainly had room for others though the relevant parts on the manuscripts have not survived.

At 76:14: Their prosperity and their interpretation I have sh[own. (4Q209 23 2)

At 74:1 (?):] another calculation I was shown[17] for it that it went (4Q209 25 3)

At 79:1: Now I am telling you, my son *blank* [(4Q209 26 6)

?]a calculation he showed m[e

Enoch claims to have seen a remarkable set of the world's features, including ones at the ends of the earth. But is anything he claims to have seen implausible? Note that he nowhere asserts he saw an angel leading heavenly luminaries; the leaders of the stars in ch. 82 are never called angels and may simply be stars.

E. Sources: A study of the astronomical booklet leads to the conclusion that Enoch's scientific teachings are based, at least in large part, on two major sources: sections of the Hebrew Bible and an early

[17] For the passive form here, see the detailed discussion in the paper of Seth Sanders in this volume and Drawnel, *The Aramaic Astronomical Book*, 314-315 (he thinks it "denotes the mental process of studying and learning the astronomical computus" [315]).

form of astronomy that comes to expression in cuneiform works such as Mul.Apin and Enuma Anu Enlil 14.[18]

1. The Hebrew Bible: If one thinks of ancient science, even of early Jewish science, the Hebrew Bible does not spring to mind as a prominent source for such material. There is in it no composition or part of a composition that could be called scientific even with an elastic definition of the term (for definitions of the term, see the discussion below). This is not to say that no one during the biblical period did scientific work; it is merely to say that nothing recorded in the Hebrew Bible is a scientific composition.[19] Qohelet could, in a sense, be an exception. The sage set up an experiment and tried to carry it out in a systematic, logical manner to arrive at a conclusion. As Michael Fox describes what he did, Qohelet used his reason to examine experience in order to produce knowledge. He wished to attain it through discovery, not simply to repackage prior knowledge. He thus went about investigating his world; his empirical argumentation proceeded from sensory experience with an emphasis on validation. He observed in order to gain knowledge and reported on his discoveries (using expressions such as "I saw," "I realized"). For him there was no independent external standard. In this way Qohelet sought to produce knowledge that did not exist before and in a sense relativized that knowledge.[20] But it would be difficult to label even Qohelet a scientific treatise.

Though it contains no scientific treatises, the Hebrew Bible does include a number of statements that could serve as foundations for the development of a scientific outlook regarding aspects of the universe.

a. Genesis: In Gen 1:1-2:4a God created an orderly universe in six days. The entire account reflects a simple classification of entities. For example, it divides between the plants and the fruit trees and separates the various kinds of beings into ones living in the waters, in the air, and on the land. Also, in its orderly approach, the creative

[18] See Annette Y. Reed, ""Was There Science in Ancient Judaism? Historical and Cross-cultural Reflections on 'Religion' and 'Science,'" *SR* 36 (2007): 467-476. She too speaks about the combination of exegetical and scientific inquiry.

[19] One could consider the classification systems in various places in the Priestly source as scientific in nature; see Sanders' paper in this volume.

[20] Michael Fox, "Qohelet's Epistemology," *HUCA* 58 (1987): 137-154.

work of God proceeds in an evolutionary manner, culminating in the fashioning of human beings, male and female. Each item in the creation has its own place and function. So the larger luminary rules the day, the smaller one the night, and humans control the rest of creation. There is no suggestion or hint that this orderly arrangement will ever change or end. A stable order is an important condition for the descriptive, classifying work of scientists, although they must also deal with what appear to be disruptions of that regularity.

The next unit in the Priestly source is Genesis 5. Its genealogy rests on a chronological system that may be keyed to the date of the flood—a system whose development was perhaps a scientific pursuit in itself. The flood, a subject that dominates chs. 6-9, complicates matters because it nearly destroyed the order established in Genesis 1. In the sequel to the story of the deluge, there are foundational statements offering information that would allow the fashioning of a scientific worldview. According to the J account, the time after the flood will be characterized in this way: "As long as the earth endures, seedtime and harvest, cold and heat, summer and winter, day and night, shall not cease" (Gen 8:22).[21] The Priestly source approaches the matter somewhat differently. In Genesis 9, where creation language is repeated, the deity promises: "I establish my covenant with you, that never again shall all flesh be cut off by the waters of a flood, and never again shall there be a flood to destroy the earth" (9:11). The stability of the creation is restored and will not again be disrupted.

It is clear from 1 Enoch 72-82 that the writer of the Enochic text knew and used Genesis 1, especially the passage regarding the creation of the heavenly lights on day four. So, for example, he designates the sun "the great light" (72:4, 35-36) and the moon "the smaller light" (73:1), and he refers to sign(s) in connection with the luminaries (72:13, 19; cf. 78:7 and 82:16, 19). Furthermore, 1 Enoch 75:3 and 82:7, 9-10 contain summary statements that take up several terms from Gen 1:14-19: times, days, years, set times, and rule.[22]

[21] Biblical quotations are from the NRSV.

[22] See James C. VanderKam, "Scripture in the Astronomical Book of Enoch," in *Things Revealed: Studies in Early Jewish and Christian Literature in Honor of Michael E. Stone* (ed. E. Chazon, D. Satran, and R. Clements; JSJSup 89; Leiden: Brill, 2004), 93-97. In the same essay there is a treatment of Isa 30:26 which

b. Other passages regarding a stable created order: The writer of the Enochic work could have based his assumption about the unchanging character of the natural order on another series of scriptural passages, although he is not as explicit about this as he is regarding his use of Genesis 1. One is Jer 31:35-36 which not only expresses the idea but also uses a term important in 1 Enoch:

> Thus says the Lord, who gives the sun for light by day and the fixed order [חקת] of the moon and the stars for light by night, who stirs up the sea so that its waves roar —the Lord of hosts is his name: if this fixed order [החקים האלה] were ever to cease from my presence, says the Lord, then also the offspring of Israel would cease to be a nation before me forever.[23]

Earlier in the book, Jeremiah used the fixed order of nature as a platform for moral judgment on his contemporaries in a way that later became popular: "I placed the sand as a boundary for the sea, a perpetual barrier [חק עולם] that it cannot pass; though the waves toss, they cannot prevail, though they roar, they cannot pass over it. But this people has a stubborn and rebellious heart ..." (5:22-23).[24] The fixity of nature to which he appeals is in line with the impression given by Genesis 1 and the post-flood statements and would be congenial to a scientific outlook.

These passages stand in contrast to ones that predict the dissolution of the natural order before the end arrives. In an interesting contrast to the Jeremiah sections just quoted, Second Isaiah was able to use the breakdown of the natural order in a positive fashion: "Lift up your eyes to the heavens,/ and look at the earth beneath;/ for the heavens will vanish like smoke,/ and the earth will

speaks about how, some day, "the light of the sun will be sevenfold" that of the moon (apparently). This is their relation according to 1 Enoch 73:3; 78:4. The verse is another important scriptural basis for parts of 1 Enoch 72-82, though it does not deal with the present order of nature (see pp. 97-103).

[23] The term חק seems to lie behind some uses of Ethiopic *śer'at*.

[24] The passage is particularly interesting in that it speaks of a "fixed order" in connection with the sea which is elsewhere treated as a threat to that order (see Robert Carroll, *Jeremiah: A Commentary* [OTL; Philadelphia: Westminster, 1986], 187). See also Job 38-41, a section where the emphasis falls on human inability to understand parts of the creation.

wear out like a garment,/ and those who live on it will die like gnats;/ but my salvation will be forever,/ and my deliverance will never be ended" (Isa 51:6). The writer of the Enochic work clearly adopted the approach in the other series of texts.[25]

2. Mesopotamian sources: This is not the place to treat the topic in detail, but, as a number of scholars have shown over the last few decades especially, the science that comes to expression in the *Astronomical Book* of Enoch is beholden to a type of astronomy attested in sources such as Mul.Apin and Enuma Anu Enlil 14. In the former, there are close parallels to Enoch's astronomy in some of the proportions (e.g., for times of light and darkness during the days in a year and in the linear progressions for the luminaries). The four tables in the latter provide interesting similarities with the lunar material in 1 Enoch. Tables A and B give data for each day of the month, and C and D cover an entire year, selecting just two dates for each month. The tables do not furnish exactly the same numbers as in Enoch's work: in them all months have 30 days and the fractions are fifteenths. But they utilize the same linear progressions and schematic form, e.g., for the time the moon is visible/invisible in the sky. The basic linear patterns are of the same type in the two works.[26]

II. The Astronomical Book of Enoch and Ancient Science

A. Science: Whether the material in the *Astronomical Book* of Enoch should be labeled *science* depends, of course, on what is meant by science. The authors of several papers in this volume have formulated

[25] In his essay "Enoch and the Beginnings of Jewish Interest in Natural Science," in this volume, Philip Alexander distinguishes two attitudes toward the physical world in the wisdom tradition of the Persian period. One comes to expression in Job 38-41 where it is assumed that the ways of God in the physical world are beyond human understanding; the other is in Proverbs 8 where, one can infer, it is a good thing for humans to study the wisdom that created the world. The circles behind the Enoch tradition took the Proverbs 8 approach that corresponds with the one adopted in the series of passages (such as Genesis 1) surveyed above, though for Alexander the scriptural passages are not scientific in his sense of the word (240; for his definition of science, see below).

[26] Detailed treatments of the subject include Matthias Albani, *Astronomie und Schöpfungsglaube: Untersuchungen zum astronomischen Henochbuch* (WMANT 68; Neukirchen: Neukirchener Verlag, 1994), 155-272; Ben-Dov, *Head of All Years*, 153-396; and Drawnel, "Moon Computation," 3-41; *The Aramaic Astronomical Book*, 301-311.

definitions of the term. So, for example, J. Ben-Dov writes that it is "the systematic observation of natural phenomena in an attempt to describe their regularity and make sense of the irregularities. This attempt involves a specialization of knowledge, as well as the use of earlier scientific corpora by way of translation or accommodation."[27] In 2002 P. Alexander, in the context of dealing with Enochic booklets, wrote that

> we can identify 'science' wherever we find a strong interest in understanding how the physical world works, provided three simple conditions are fulfilled: (1) There is an explicit or implicit assumption that nature is regular and is governed by immutable laws which are accessible to the human mind. (2) An attempt is made to produce a rational model of the physical world which reduces the bewildering complexities of natural phenomena to a small number of underlying primary elements, or to the operation of a small number of fundamental laws. (3) Explicitly or implicitly, a significant element of direct observation of the physical world is involved.[28]

Adopting aspects of definitions such as these, it can be seen that the material in 1 Enoch 72-82 may be called science.

1. Systematic observation of natural phenomena that are assumed to operate according to consistent laws accessible to human understanding: As noted above, a presupposition of Enochic thought is that God has assigned the various parts of the created world specific laws that they obey without exception (unlike people). The patterns for the solar and lunar years do not change, as sun, moon, and stars always travel upon their assigned paths at the times prescribed for them. Those patterns Enoch learned from observation of phenomena and perhaps from instruction in their patterns—all directed by the

[27] See his paper in the volume.
[28] "Enoch and the Beginnings of Jewish Interest in Natural Science," 224.

angel Uriel.[29] That is, the correct information is presented as being accessible to a human being despite the fact that he obtained it in a special way. The luminaries operate in a consistent fashion that can be described and charted. The writer also reduced something as complex as lunar motion to simple, obviously overly simple, patterns; he apparently did the same for the sun.

2. Use of earlier scientific corpora: Again as indicated above, the material in the Enochic work is based on more ancient sources containing similar information (especially the more primitive kind of astronomy in Mul.Apin and Enuma Anu Enlil 14, but with influence from scriptural texts), although the writer developed the inherited evidence in accord with his understanding of the way the world worked. It is reasonable to suppose he drew such information from works written in Aramaic, the language in which he himself was

[29] Revelation of scientific data is not a misnomer in an ancient text, as we have seen. Alexander thinks the writers responsible for Enoch's astronomical work appealed to revelation because "they were consciously attempting to domesticate within Jewish tradition a body of alien wisdom" ("Enoch and the Beginnings of Jewish Interest in Natural Science," 232). An appreciable part of Enoch's science derives from foreign sources, but why should such borrowing require a special literary framework—revelation by an angel—to "domesticate" it? Does Genesis resort to such a framework to introduce creation traditions or flood stories drawn from non-Israelite sources? The revelatory basis in the *Astronomical Book* is related to the exegetical tradition that Enoch spent time with the angels (Gen 5:21-24) and the further development that a being with a name like Uriel would be the one associated with information about the workings of the celestial lights. One should also object to Alexander's proposal that an Enoch-Moses rivalry lies behind the revelatory framework—Enoch is associated with "a paradigm based primarily on science as opposed to one based primarily on law" for Moses (234). One can infer from some texts that Moses did not enjoy as high a status in the Enochic tradition as he did elsewhere (e.g., the Apocalypse of Weeks [1 Enoch 93, 91]), but if we confine our attention to the astronomical writing of Enoch it is inappropriate to claim that such a rivalry is present in the sense that the Enochic circle was trying to establish a new, scientific paradigm as opposed to one based on law. In Enoch's book about the luminaries the focus is on the scientific evidence and nothing is said about covenant and the like. Whatever may have happened later, there is no indication in it of the conflict that Alexander describes. I do not know whether the Enochic people were trying to introduce a new scientific paradigm or were merely attempting to offer a fuller explanation of phenomena than Genesis and other parts of the Bible provide. That fuller explanation need have no implications for a worldview based on the Mosaic law. Alexander, it should be said, has in mind more than the *Astronomical Book* when he writes about a Moses-Enoch rivalry.

writing.[30] I believe the astronomical booklet could have been written in the eastern diaspora where there would have been direct contact with the astronomical traditions known to us from Mul.Apin and Enuma Anu Enlil 14.[31]

3. Based on observation: One of the pervasive elements in 1 Enoch 72-82 and one attested in the Aramaic fragments as well is that Enoch saw or was shown the phenomena about which he wrote (see above). At any rate, that is how it is presented in the text.

B. Purpose: All of this raises the question why the author wrote the *Astronomical Book* of Enoch. One reasonable suggestion is that priests would have had need for an astronomical compendium of information for calculating the times for festivals, new moons, and similar dates, although festivals are rarely mentioned in the booklet. As Drawnel has argued, the work could fall into the category of didactic priestly instruction:

> The comparison between the *Aramaic Levi Document* and the *Aramaic Astronomical Book* helps explain the fragmentary text of the latter composition. The astronomical text that intends to calculate monthly moon illumination belongs to the priestly lore of didactic literature, in which simple arithmetical knowledge was used both for the sacrificial purposes and astronomical calculation. The didactic character of the *Astronomical Book* appears in the literary pattern of knowledge transmission according to which father/-teacher instructs his son/pupil and the vocabulary pattern of "showing" and "seeing" of the learned subject

[30] Alexander quite understandably proposes that the well-educated circles behind Enochic astronomy belonged to the scribal and priestly classes of society ("Enoch and the Beginnings," 239).

[31] See Albani, *Astronomie und Schöpfungsglaube*, 248-272; VanderKam, *1 Enoch 2*, 383.

> metaphorically refers to the didactic role of the teacher and student respectively.[32]

Perhaps that is how we should view this science: as an attempt to produce something very schematic to provide basic knowledge to aspiring priests or others whose work would require some knowledge about the workings of God's creation, especially the ways to measure time.[33]

[32] Henryk Drawnel, "Priestly Education in the *Aramaic Levi Document* (*Visions of Levi*) and *Aramaic Astronomical Book* (4Q208-211)", *RQ* 22/88 (2006): 561-562; cf. 567. He adds that he finds it reasonable to think the booklet "was composed by a priestly teacher who intended to pass to his priestly students simple astronomical and calendaric notions vested in the garb of an angelic instruction" (567). Such practices parallel and were influenced by Babylonian models. His argument that the didactic context would also lead one to expect a section such as 1 Enoch 80:2-8 misses the point that these verses do not talk simply about a misunderstanding of astronomical laws by sinners but an actual breakdown in the natural order (see 562-565).

[33] See also VanderKam, *1 Enoch 2*, 359, 367-368.

4. "I Was Shown Another Calculation" (חשבון אחרן אחזית): The Language of Knowledge in Aramaic Enoch and Priestly Hebrew[1]

Seth L. Sanders

> The science of apocalypticism can be defined as the exact numerical calculation of the end of time. It is intended to provide absolute assurance to faith and hope. The science of apocalypticism, which numerically calculates the *when* of the End Time, rests on the belief that "everything must fulfill its course out of inner necessity." It is the task of the seer to reveal this necessity. —Jacob Taubes[2]

The earliest known Jewish scientific work is, probably not coincidentally, also the first known scientific work in Aramaic. This is the *Astronomical Book* of Enoch, found at Qumran, the oldest manuscripts of which date to the 3rd century BCE The text is written in the cosmopolitan, high-cultural register of the *lingua franca* of the Babylonian and Persian empires known as Standard Literary Aramaic.[3]

[1] This paper was originally presented at the "Ancient Jewish Sciences and the History of Knowledge" conference at the NYU Institute for the Study of the Ancient World on April 4, 2011, where Loren Stuckenbruck delivered a valuable response, which follows the paper. The ideas emerged from a discussion in the hallway of ISAW with my colleague and co-organizer Jonathan Ben-Dov; I thank him for introducing me to this remarkable set of issues. This draft was improved by detailed comments from the viewpoints of biblical and Second Temple literature by Ben-Dov, of Hebrew and Aramaic linguistics by Edward Cook and Matthew Morgenstern, and the history and philosophy of science by Michael Barany. It has also benefitted from an inspiring discussion with Simeon Chavel and valuable remarks by Kelley Coblentz-Bautch, Daniel Stökl Ben-Ezra, and Tzemah Yoreh. All errors remain my own.

[2] *Occidental Eschatology* (trans. David Ratmoko; Stanford: Stanford University Press, 2009 [1947]), 32. Note that the original German term Taubes used was *Wissenschaft,* which has a broader range than English "science," denoting any form of rigorous scholarship.

[3] For an incisive discussion of linguistic variation in the Aramaic found at Qumran and what it may say about its associated textual genres and producers, see Aaron Koller, "Four Dimensions of Linguistic Variation:

Its contents consist of a series of rules for the movement of the heavenly bodies and the increase and decrease of hours of light over the course of the year, presented as visions seen by the antediluvian patriarch Enoch during a tour of the universe conducted by the angel Uriel.

But why is the oldest known Aramaic science presented to a Jewish patriarch during a heavenly journey? Babylonian mathematics and astronomy are generally considered by historians of science to be the most highly developed empirical knowledge in the ancient world. This paper attempts to shed some light on the peculiar way that this most exact of sciences was presented in the Levant: as a set of mystical visions in an incipient Jewish apocalyptic literature.[4]

The paper will approach this broad problem narrowly: by placing the *Astronomical Book* in the cultural context of Hellenistic Judea and asking what one of the text's key linguistic patterns—one heretofore little-recognized—tells us about this context. I take as an axiom that the role of social theory in philology is to frame interesting questions that the data are suited to answer. In this case, to understand what type of knowledge the *Astronomical Book* contains, it may not be as useful to ask "is it science for us?" as it would be to ask, "was it science for them?" and "if so, how?"

These questions set my agenda: to avoid anachronism and place the *Astronomical Book* in the history of science, we need to ask what systematic knowledge of the physical world might have meant to Hellenistic Jewish writers. The inherited genres of exact knowledge that are both well preserved to us and likely to have been important to

Aramaic in and around Qumran" in *The Dead Sea Scrolls in Context: Integrating the Dead Sea Scrolls in the Study of Ancient Texts, Languages, and Cultures* (ed. Armin Lange et al.; SVT 140; Leiden: Brill, 2011) I:199-213, with references. The foundational study was by Jonas Greenfield, "Standard Literary Aramaic" in *Actes du premier congrès international de linguistique sémitique et chamito-sémitique, Paris 16-19 juillet 1969* (ed. A. Caquot and D. Cohen; The Hague: Mouton, 1974), 281-289; repr. in *'Al Kanfei Yonah: Collected Studies of Jonas C. Greenfield on Semitic Philology* (ed. S.M. Paul, M.E. Stone, and A. Pinnick; 2 vols.; Jerusalem: Magnes, 2001), 1:211-220.

[4] For detailed discussion of the nature of the scientific material in Enochic literature see the essays of Alexander and VanderKam in this volume. The Enochic *Book of Watchers* is taken to be paradigmatic of early Jewish mystical literature even by the most skeptical student of the topic, Peter Schäfer, for which see his *Origins of Jewish Mysticism* (Tübingen: Mohr Siebeck, 2009).

scholars in Hellenistic Judea are contained in the Priestly literature of the Pentateuch, so the first half of the paper will examine how these texts organize knowledge of the physical world. In the second half I will examine an Aramaic phrase that introduces formulae for calculating the hours of light in *Astronomical Enoch*, as well as cosmic geography in the Enochic *Book of Watchers*. Although it has not been remarked before, this phrase shows striking signs of being calqued on the Hebrew phrase that introduces plans for the tabernacle in Exodus. If so, the language of knowledge in Enochic literature is closely based on the language of knowledge in Priestly literature.

This linguistic and literary pattern can cast some new light on how "science" was framed by early Jewish scholars. Like its Babylonian ancestor, Enochic astronomy claims to be based on the observation and calculation of originally divine phenomena. But for Enoch, these Babylonian calculations are presented through a story in which the data as well as their analysis are divinely revealed. The revelation of astronomy and cosmic geography to Enoch is based on the revelation of the tabernacle to Moses in Exodus. Just as Moses is "caused to see" the proportions of the temple, Enoch is "caused to see" the calculations that specify the movements of the spheres. By emphasizing Enoch's passivity in the observation and calculation of heavenly phenomena, these texts frame what we would consider empirical knowledge of the physical world in a very different kind of epistemology: a "revealed science" that helps explain how astronomy could function as part of an apocalypse.

1. Did They Believe in Science? Ancient Jewish Knowledge about the Physical World

Biblical texts show no interest in mathematics, astronomy, or what we would call science. Indeed, Deuteronomy 4:19 contains an explicit and pointed warning *against* the dangers of visual evidence of the physical universe. Observation of the heavenly bodies is a temptation towards, perhaps because it provides evidence of, other gods—and therefore forbidden to Israel.[5]

[5] The evidential language of the passage is remarkable: it reads ופן-תשא עיניך השמימה וראית את-השמש ואת-הירח ואת-הכוכבים כל צבא השמים ונדחת והשתחוית להם ועבדתם אשר חלק יהוה אלהיך אתם לכל העמים

But already before the completion of the Hebrew Bible, some Jewish writers were adopting very different attitudes toward knowledge of the physical world. Even as books like Daniel were being completed, narrating the uselessness of foreign knowledge in comparison to the revelations of Israel's God (e.g. Dan 1:20, 2:19), Jewish writers had begun incorporating Babylonian mathematics and

תחת כל-השמים "Lest you look up at the heavens and see the sun, the moon, and the stars—the whole entourage of heaven—and become scattered by bowing down to them and worshiping them, who the Lord your god assigned to all the (other) peoples under heaven." (Note that all translations of Hebrew and Aramaic texts are mine.) In a move strikingly relevant to the present discussion, but which cannot be pursued here, Steven Weitzman argues that this is part of a distinctive Deuteronomistic agenda for disciplining the senses. In this agenda, memory—but not visual evidence—is to be relied on for religious knowledge and practice. See his "Sensory Reform in Deuteronomy" in *Religion and the Self in Antiquity* (ed. David Brakke et al.; Bloomington: Indiana University Press, 2005), 123-139, esp. 128-9. Weitzman's arguments complicate the more standard understanding of the passage as "express[ing] clearly the fact that from biblical times on, Jewish faith has been based primarily on experience rather than speculative thought" (Jeffrey Tigay, *Deuteronomy* [Philadelphia: Jewish Publication Society, 1996] ad 4:9, with citation of A. J. Heschel). Weitzman's arguments can be strengthened by a close reading of the passage in which the line appears, Deut 4:12-19. What is distinctive about the passage is its repeated stress on the nature of experience, and what claims follow from that sort of experience. The experience was, first of all, *auditory*: 4:12 and 4:15 state: "The Lord spoke to you out of the fire, and it was the sound of words you heard—you saw no image, only a sound. ... So keep ... watch over yourselves, since you saw no image when the Lord spoke to you out of the fire at Horeb ..." Then the text warns of two dangerous misapprehensions, each involving acting on the evidence of one's vision. Formally, both are expressed by "lest" plus a pair of subjunctive verbs joined hendiadically: "Lest you cause ruin by making (פן-תשחתון ועשיתם) for yourselves a sculpted image ..." (4:16) and "Lest you look up to the heavens and see" the heavenly bodies "and become scattered by bowing down (ונדחת והשתחוית) to them and worshiping them ..." (4:19). The first pathway to disaster is making a visible image and taking it as a divine being; the second pathway is seeing the visible images of the stars and taking them for divine beings. Now, Mesopotamian astronomy and astrology were based on the idea that the stars were in fact the visual evidence of gods, "heavenly writing," in the Babylonian phrase (discussed by Rochberg in her book of the same name), in which a divine pattern could be read. In response to this, the Deuteronomic passage argues not that there is no writing in the heavens, but that it is dangerious to read. It is dangerous to read one's visual experience of the sky, precisely because signs of the divine *may* be found there: evidence of more than one god. Contrast the account I give below of Priestly exact knowledge, in which visual evidence of both the cosmos and human body are crucial.

astronomy into major literary works.[6] By the third century BCE, Biblical patriarchs like Levi and Enoch were represented as learning

6 Contra Hans-Peter Müller, the text of Daniel is emphatic in contrasting the reliable knowledge revealed by God with the Mesopotamian court sages' unreliable arts of knowledge (to use a term brought into biblical studies by Esther J. Hamori; see her *Women's Divination in Biblical Literature: Prophecy, Necromancy, and Other Arts of Knowledge* f/c in the Yale Anchor Bible Reference Library). While it has been argued that Daniel's techniques have a Mesopotamian provenance, the narrative of Daniel typically states the opposite, energetically opposing his knowledge to that of the court sages—with one exception. In an important treatment, Alan Lenzi has emphasized the contrast between Daniel, who receives revelation directly from God, and the Babylonian scholars who relies on written sources and sensory evidence, thus implicitly constrasting his successful "charismatic" performance with the scholars' failed "institutionalized" practice. While generally convincing, Lenzi's argument stumbles in forcing Daniel into a binary relationship with the Babylonian scholars; Daniel is actually presented as occupying multiple positions with respect to Mesopotamian scholarship that vary from insider to outsider. In fact, while Daniel is compared favorably to the Babylonian court scholars in Dan 1:19-20; 2:48; 4:8; 5:29; 6:28, he is only appointed over them at the end of Dan 2. By contrast, in Dan 4 he is introduced with both a Babylonian name and professional title: "Belteshazar, chief dream-interpreter" בלטשאצר רב חרטמיא (Dan 4:6), a designation that corresponds to a more elaborate title in the parallel version underlying the Old Greek. Here he is given his Jewish name but termed "chief of the sages and leader of the dream-interpreters" τὸν ἄρχοντα τῶν σοφιστῶν καὶ τὸν ἡγούμενον τῶν κρινόντων (4:15 in the Göttingen edition; for a useful treatment of the Old Greek and its relationship to the MT of Dan 4, see John Collins, *Daniel: A Commentary on the Book of Daniel* [Minneapolis: Fortress, 1993]). Dan 4 is the only part of the book to designate Daniel as a *Babylonian* expert from the outset, an anomaly most simply explained by its source, a Babylonian Aramaic legend of which an earlier version is preserved in 4Q242, the Prayer of Nabonidus. Here the figure who rescues the Babylonian king is simply termed "a diviner, a Jew" גזר והוא יהודי (1-3 4). See Lenzi, "Secrecy, Textual Legitimation, and Intercultural Polemics in the Book of Daniel" *CBQ* 71 (2009): 330-348. For the category of "Mantic (or Magic-mantic) wisdom" see Müller, "Magisch-mantische Weisheit und die Gestalt Daniels," *Ugarit-Forschungen* 1 (1969): 79-94, "Mantische Weisheit und Apokalyptik," in *Congress volume: Uppsala, 1971* (SVT 22; Leiden: Brill, 1971): 268-293. James VanderKam insightfully presents and analyzes the Qumran evidence in his "Mantic Wisdom in the Dead Sea Scrolls," *DSD* 4 (1997): 336-353. But the gap between Daniel's revealed wisdom and the Mesopotamian arts of knowledge raises questions for the category of "mantic wisdom": what makes divination a category of wisdom? The two genres are distinct in Mesopotamian and biblical literature, and the features VanderKam analyzes in the scrolls do not fit neatly with the generic features of wisdom recently identified by Matthew Goff, "Qumran Wisdom Literature and the Problem of Genre," *DSD* 17 (2010): 286-306. cf. the brief critique of Andreas Bedenbender, "Jewish Apocalypticism: A Child of Mantic Wisdom?," *Henoch* 24 (2002): 89–96.

and teaching about numbers and the stars. And rather than crude improvisations, these teachings derived from the highly developed techniques of Babylonian scholarship, arguably the world's first truly empirical scientific tradition.[7] By the first century BCE, these new interests had been energetically and creatively integrated into the ritual framework of the Qumran community.[8] And these texts are only the earliest evidence of a pattern of systematic cosmological speculation in Jewish tradition, often presented as exegesis of Genesis 1, that continued to assume new forms through the Byzantine and medieval periods.[9]

[7] For the empirical basis of Babylonian astronomy and its importance in the history of science, see Noel Swerdlow, *The Babylonian Theory of the Planets* (Princeton, N.J.: Princeton University Press, 1998). For its historical context and its role in debates about the nature of science see Francesca Rochberg *The Heavenly Writing: Divination, Horoscopy, and Astronomy in Mesopotamian Culture* (Cambridge, UK and New York: Cambridge University Press, 2004). Recent decades of research on the texts from Qumran have revealed that the earliest known Jewish apocalypse, the *Astronomical Book* of Enoch, is also the earliest known piece of Jewish astronomy and detailed mathematical calculation. Paleographic dating of the manuscripts place this text's *Vorlage* in the third century BCE or earlier, before the final form of biblical books like Daniel. On the development of the books and figure of Enoch see James VanderKam, *Enoch and the Growth of an Apocalyptic Tradition* (Washington, D.C: Catholic Biblical Association of America, 1984). The thorough and persuasive treatment of the Babylonian mathematics and astronomy in the *Astronomical Book* and at Qumran by Jonathan Ben-Dov, *Head of All Years: Astronomy and Calendars at Qumran in their Ancient Context* (STDJ 78; Leiden: Brill, 2008), builds on but supersedes previous work on the subject. For the dating of the *Aramaic Levi Document* and the provenance of its measurement system see Henryk Drawnel, *An Aramaic Wisdom Text from Qumran. A New Interpretation of the Levi Document* (JSJSup 86; Leiden: Bril, 2004).

[8] Ben-Dov has recently argued that the mid-second-century BCE Hebrew text 4Q317 "is in fact a translation and adaptation of the oldest section from the Aramaic *Astronomical Book*, attested in 4Q208 and 4Q209," and the same concepts were later adapted and integrated into the sectarian Mishmarot texts see Ben-Dov, "Scientific Writings in Aramaic and Hebrew at Qumran: Translation and Concealment" In *Aramaica Qumranica: Proceedings of the Conference on the Aramaic Texts from Qumran in Aix-en-Provence, 30 June-2 July 2008* (ed. K. Berthelot and D. Stökl Ben Ezra; STDJ 94; Leiden: Brill, 2010), 393, and on the adaption of EnAstr 394; and Ben-Dov, *Head of All Years*, 147–151 on the adaption of concepts underlying 4Q317.

[9] Tensions—some parallel to the ones treated here—exist in the major medieval Jewish esoteric sources on the role of the stars. These are treated incisively by Ronald Kiener, "Astrology in Jewish Mysticism from the Sefer Yezira to the Zohar," *Jerusalem Studies in Jewish Thought* VI (3-4) (1987): 1*-42*. For the categories of cosmic knowledge from Hellenistic through

Was there a Jewish "apocalyptic science," as Jacob Taubes boldly proposed in 1947? These cases of serious interest in mathematics and astronomy raises the question of whether the Hellenistic period represented the dawn of a kind of scientific thought in Judaism. It is clear that *something* new was dawning, for which current historical frameworks are not quite adequate.

For example, while Hellenization is a common way of how new ideas emerged in early Jewish literature, it is not possible to explain the new Jewish interest in mathematics and astronomy through Greek influence. This is because it mainly draws on elements that had existed in the ancient Near East for centuries before its emergence in texts of the third century BCE. The new material first appears in Aramaic, not Greek; it is in a dialect—Standard Literary Aramaic—formed already in the Persian period; and it derives directly from a Babylonian scientific tradition that was itself one of the main influences on early Greek mathematics and astronomy.[10] Since this new phase of Jewish thought cannot be explained as a result of the Hellenization that has been so central to our explanations of other major changes in Judaism, from the rise of Rabbinic Judaism to Christianity, we need to seek other explanations.

Taubes argued for the idea of an "apocalyptic science" in a German dissertation that was more or less neglected in the study of early Judaism and Hellenistic history. But in the past decade, the idea

Byzantine Judaism see Annette Yoshiko Reed, "Was there Science in Ancient Judaism? Historical and Cross-Cultural Reflections on 'Religion' and 'Science,'" *SR* 36 (2007): 461-495 and her contribution to this volume. For the corresponding Christian transformation of ancient Hellenistic *scientia* see Hervé Inglebert *Interpretatio Christiana: Les mutations des savoirs (cosmographie, géographie, ethnographie, histoire) dans l'Antiquité chrétienne (30-630 après J.C.)* (Collection des Études Augustiennes; Paris: Institute d'Études Augustiniennes, 2001). Inglebert analyzes a series of four *savoirs*—cosmography, geography, ethnography, and historiography—that are only partly comparable to the "sciences" under discussion here. But given the strong interrelationship that Enoch's editors and later apocalyptic thinkers saw between cosmography, geography, and historiography, might it be fruitful to pursue Inglebert's series further?

[10] I owe this point to Simeon Chavel (pers. comm. 2011), who emphasizes the fact that in Jewish sources the adapted Babylonian materials are never presented as having a foreign cultural provenance. Thus not only "science" but also "Hellenistic" and "Babylonian" are in important ways reified and anachronistic terms that only somewhat awkwardly fit our data.

has reemerged as scholars begun to suggest that a new form of Jewish thought appeared in the Enochic literature. Philip Alexander was the first to argue in detail that "the approach to nature displayed in the Enochic *Book of the Heavenly Luminaries* is unprecedented in Jewish literature. It seems to mark a turning-point in Jewish intellectual history—the emergence, for the first time, of what might properly be called a scientific attitude."[11]

Alexander's argument reopened the discussion after Taubes, challenging us to ask new questions appropriate to our data. He has been followed by Jonathan Ben-Dov and Annette Reed, who have each added significant arguments for integrating early Jewish discourse about numbers, the body, and the stars into the history of science.[12] As the great Otto Neugebauer pointed out, ancient science is especially useful to the cultural historian because it often used highly distinctive, originally Babylonian methods, the trajectory of which can be traced across languages and cultures.[13] For example, the Babylonian cuneiform writing system used a base-60 number system (in contrast to the Hebrew and Greek base-10 system) and a "linear zigzag function" for calculating the length of the day (in which the longest day of the year is assumed to have a 2:1 proportion of hours of light to hours of darkness, and the shortest day is conversely assumed to have a 1:2 proportion). Because these conventions are both distinctive and arbitrary, they are easily recognizable across languages and texts and give us reliable evidence for the origins and direction of intercultural contact. Ben-Dov used the linear zigzag function, among a number of features, to deduce the precise nature and time period of the Babylonian astronomy used in the *Astronomical Book* of Enoch, and Drawnel was similarly able to show that the base-60 calculations in the *Aramaic Levi Document* had to have a Babylonian origin. The patterns shared between Babylonian, Jewish and later Greek texts pay a further dividend because they force us to rethink old and perhaps inappropriate analytical categories. As we shall see, the astronomical

[11] "Enoch and the Beginnings of Jewish Interest in Natural Science" in this volume.

[12] See Ben-Dov, "Scientific Writings"; Reed, "Was there Science?".

[13] Neugebauer, *The Exact Sciences in Antiquity* (Providence: Brown University Press, 1957, 2nd ed.).

material in Enoch cannot be categorized as "law," "prophecy," "wisdom" or "religion."

But is "science" the best category for organizing this body of data? Along with a lucid analysis, Alexander provided his own definition of science. Reed and Ben-Dov also proposed brief characterizations, each differing from the other.[14] And this points to a problem. As we show in the introduction to this volume, when one compares the divergent reasons scholars have given for why texts like the *Astronomical Book* should be called science, it is easier to agree that it is science than to specify why.

The new types of knowledge that emerged in Hellenistic Jewish culture were heterogenous. They included astronomical calculations of the movements of the heavenly bodies and length of the days, sexagesimal (base-60) mathematics, and physiognomic interpretation of the body. On the one hand, all of these modes of knowledge have at some point in modern European history been understood as natural science: astronomy and mathematics are of course still understood this way. But as late as the mid-19th-century a form of physiognomy known as "phrenology" was still taken seriously by scholars across Europe.[15] It seems intuitively correct to us to define mathematics and astronomy as exact science, but is it science to observe someone's hair to predict their character and destiny, as the Qumran physiognomic text 4Q186 does?[16] What is surprising is how clear the verdict of the history and philosophy of science is on this point: there is simply no rigorous way to tell whether a discipline is science or not.

[14] For detailed citation and analysis, see the introduction, where we argue that Alexander's definition is not compatible with some important and widely recognized features of ancient science, such as appear in Aristotle's *Posterior Analytics*.

[15] On the lives and deaths of phrenology and related physical and quantitative approaches to human character, see the lively study of Stephen Jay Gould, *The Mismeasure of Man* (New York: Norton, 1996, Rev. and expanded ed.).

[16] An illustrative passage comes from fragment 1 column ii lines 3-8: "[And] anyone [whose] eyes are [... and lo]ng, but th[e]y are fix[e]d, whose thighs are long and slender, whose toes are slender and long, and who was born during the second phase of the moon: he possesses a spirit with six parts light, but three parts in the House of Darkness. This is the birth sign under which such a person shall be born: the haunch of Taurus. He will be poor. This is his animal: the bull."

In an influential 1983 article, the philosopher of science Larry Laudan explained that the problem of distinguishing scientific knowledge from other types has loomed large in Western philosophy for a long time: "From Plato to Popper, philosophers have sought to identify those epistemic features which mark off science from other sorts of belief and activity."[17] In the philosophy of science, the task of defining the boundaries of science became known as the "demarcation problem," and after well over a century of heated debate it is now generally agreed to be insoluble. "... it is probably fair to say that there is no demarcation line between science and non-science, or between science and pseudo-science, which would win assent from a majority of philosophers" (112).

Historically, nobody seems to have ever produced a necessary and sufficient definition of science (or its ancestors such as Aristotle's *episteme*). As Laudan explains, definitions have at some points focused on science as proceeding deductively from *a priori* principles and at other points as proceeding inductively from observed phenomena. Science has also been defined as being "falsifiable" (a definition that includes the flat earth theory) or as proceeding from a "scientific method" (the rules of which were never successfully explained). And as Francesca Rochberg has shown, ancient science was also defined in contradictory ways. It has been described as explanation without accurate observation (Greek), or accurate observation without explanation (Mesopotamia and Egypt).[18]

The problem is not that no useful criteria for science are possible, but that historically, there have been so many of them. These criteria concern something immensely important to the people who proposed them—the nature of reliable knowledge of the world. So while it may be ironic that the question, "is it science?" does not admit of a "scientific" (or at least universally valid) answer, it is precisely the persistence of the question that can provide a useful starting point.

[17] "The Demise of the Demarcation Problem," in *Physics, Philosophy, and Psychoanalysis: Essays in Honor of Adolf Grünbaum* (ed. R.S. Cohan and L. Laudan; Boston Studies in the Philosophy of Science 76; Dordrecht: D. Reidel, 1983), 111.

[18] Rochberg, *Heavenly Writing*; cf. the statement of Feyerabend she cites on 31 and contrast the dichotomy of Babylonian as quantitative but not theoretical, vs. (later) Greek as both quantitative *and* theoretical cited on 34.

Rather than trying to place our texts into an anachronistic modern category, we must first find out how biblical and early Jewish writers themselves depicted systematic knowledge of the physical world.

The following investigation will adopt a provisional, heuristic definition: "science" will be understood as a system of exact knowledge of the physical world. This will let us compare a usefully large, but delimited, group of corpora. It will let us investigate what counted as reliable knowledge of the physical world for the ancient Jewish writers of Enoch: did they presuppose the Classical opposition between *physis* "nature" and *nomos* "culture," i.e. what is physically given in the cosmos and body, versus humanly given norms? If they did not presuppose such oppositions, then what, if anything, separated their astronomy from other forms of systematic technical knowledge such as law? Similarly, we will ask how their knowledge of the physical world was opposed or related to categories usually associated with religion, such as ritual, for example in the forms of knowledge laid out in Priestly texts and later Qumran works of *halakha*. Did our writers differentiate between cosmic and ritual knowledge, rather than claiming, for example, that God created a seamless network of nature and culture? As we shall see, Enoch's authors inherited a set of texts that did not assume an essential difference between nature and culture, but rather work on a set of homologies between what is created and what is commanded.

Finally, this intellectual shift will only acquire its full meaning in a historical and cultural context, but we must move forward incrementally in reconstructing one, attentive to what the sources are best suited to tell us. We still lack first-hand information about the people who produced the early Enochic literature. In response, scholars have so far reconstruct a cultural and historical setting based on attitudes and allusions reflected in the texts.[19]

[19] Second Temple scholars often resort to reading groups of literary texts as if they stood for social groups. Representative of the problem is the question of whether the Enochic literature, in contrast to Pentateuchal literature, was the product of an "Enochic movement" separate from a mainstream "Mosaic" Judaism. The most prominent contemporary proponent of an Enochic break with earlier Priestly traditions in Judaism is Gabriele Boccaccini, who lays out his thesis in *Beyond the Essene Hypothesis: The Parting of the Ways between Qumran and Enochic Judaism* (Grand Rapids, Mi.: Eerdmans,

What we can be sure of is that these early Jewish writers inherited a set of biblical categories, but transformed them to accommodate Babylonian science.[20] Their scientific sources emerged from an old cosmopolitan Mesopotamian world of exact descriptive knowledge, one that had existed for centuries in cuneiform, and then Aramaic, independent of Priestly and Deuteronomistic categories.[21] Since there are no biblical genres of mathematics or astronomy, and divination is taboo, the extensive use of Babylonian astral science in Aramaic texts at Qumran could potentially appear as a radical break within Judaism.[22]

1998). Veronika Bachmann provides a useful summary of the state of the question in "The *Book of Watchers* (I Enoch 1-36): An Anti-Mosaic, Non-Mosaic, or even Pro-Mosaic Writing?," *JHS* 11/4 (2011), online at http://www.jhsonline.org/Articles/article_151.pdf. The title of her article accurately represents the ambiguity of the current evidence.

[20] If so, did it create a rupture with long-term Judean/Jewish/Hebrew traditions? Note well that these three adjectives do not form a smoothly overlapping series—surely part of the point, and perhaps part of the solution? As Michael Stone points out in his response to Ben-Dov's article, "Émile Puech [argues that] the linguistic milieu of Qumran was no different from that of the rest of contemporary Judea. I propose considering that the same is true of the "scientific" milieu. In fact, we have very little information about the greater culture in which the Jews in the land of Israel lived, either in the First or Second Temple periods. If we were dependent on the Hebrew Bible, virtually nothing, for the Hebrew Bible does not deal with scientific issues ... All considered, however, it is probable that the 'larger culture' in which the Jews lived was basically Mesopotamian." Stone *apud* Ben-Dov, "Scientific Writings in Aramaic and Hebrew at Qumran," 400.

[21] For the intertwining of the Babylonian and Aramaic script-languages and intellectual worlds see the rich presentation of Paul-Alain Beaulieu, "Official and Vernacular Languages: The Shifting Sands of Imperial and Cultural Identities in First-Millennium B.C. Mesopotamia" in *Margins of Writing, Origins of Cultures* (ed. Seth L. Sanders; OIS 2; Chicago: Oriental Institute, 2006), 187–216.

[22] While biblical texts condemn divination, early Enochic literature presents multiple positions. Significant Deuteronomistic prohibitions of divination appear in Dt 18:10, 14; 1 Sam 15:23; 2 Kings 17:17. A remarkable scene in Ezekiel depicts the king of Babylon as employing a grab-bag of divination techniques that help him decide to attack Jerusalem (21:26-28, where the root *qsm* appears four times). By contrast, Enoch presents contradictory positions on the study of the heavenly bodies, as an impetus to both obdedience and disobedience (1 En 2) and blasphemy (1 En 7-9) in the editorially complex *Book of Watchers*. Annette Reed explores this tension: Enoch promotes astronomy in 1 En 2 and the *Astronomical Book*, but it is condemned in the three accounts of the angels' fall in 1 En 7-9, where transmitting knowledge of the workings of the stars and the course of the

Yet when Babylonian material appears in Enoch it seems to occasion no break at all; it is presented in terms of the inherited biblical categories of knowledge. This itself is a valuable clue to the cultural world in which our writers lived. We do not know the full range of texts that Second Temple Jewish scribes had available to them.[23] But biblical texts are the most solid ground on which we can begin a description of their intellectual framework. We can be certain that they were aware of biblical texts containing exact descriptions of the cosmos, temple, and human phenomena. It is to these texts that we now turn to provide some of the background for Enoch's new knowledge.

2. Created and Commanded, Nature and Culture: Priestly Categories and their Legacy

Because they provide the most exact chronological framework of the Pentateuch and contain the most extended discussions of the physical world, Priestly texts are a reliable starting point for considering the scientific background of *Astronomical Enoch*. As knowledge of the physical world, the material in Enoch would have been understood in light of three major corpora, concerning 1) time and the universe, 2) the temple, and 3) the human body. These are contained in the creation account of Genesis 1-2:4a, the temple revelation of Exodus 25-31 (cf. Ex 35-40 and Ezekiel 40-48), and Leviticus 12-15, with its command to observe physical signs to diagnose them, as it were, as symptoms of a ritual state, primarily the form of impurity manifested in the disease צרעת.[24]

moon, as well as the use of plants for healing, constituted a major part of the fallen angels' transgression. See Annette Y. Reed, *Fallen Angels and the History of Judaism and Christianity: The Reception of Enochic Literature* (New York: Cambridge University Press, 2005), 37-44.

[23] As Michael Stone remarked about the Qumran Enoch fragments, "In principle, there is no reason to think that the body of literature that is transmitted as the Hebrew Bible is a representative collection of all types of Jewish literary creativity down to the fourth century." See "The Book of Enoch and Judaism in the Third Century B.C.E.," *CBQ* 40 (1978): 490.

[24] Any thorough exploration of this topic will also need to examine continuities with the geographical and cosmographical lists discussed most perceptively by Michael Stone, "Lists of Revealed Things in the Apocalyptic Literature" in *Magnalia Dei, the Mighty Acts of God: Essays on the Bible and archaeology in memory of G. Ernest Wright* (ed. Frank Moore Cross, Werner

The Hebrew Bible begins with the Priestly account of the origins of the physical world, a pointedly taxonomic narrative in which each major sort of thing in the world is created, category by category.[25] This creation account ends by narrating how a seven-day ritual week is built into the structure of the cosmos (Gen 2:2-3). It is God's speech that performatively completes the cosmos with a verbally sanctified cycle of seven days, which fact is transmitted by an anonymous Priestly author as the definitive account of creation. Similarly, in a ritual text which appears to have been shared by both the Deuteronomistic and Priestly schools (Lev 11, Dtr 14), the prohibition on eating creatures derives directly from their observable physical characteristics: the category of "unclean" (טמא) completely overlaps with the category of "abominable" (תועבה in D)[26] and therefore prohibited in Dtr 14:3-20. In both cases, Priestly texts present the pattern of divine *commands* as homologous with the pattern of divine *creation*.

The idea that animals have different inherent types of physical nature is broadly-based in the Hebrew Bible, appearing in narrative as well as ritual sources. In addition to D it appears not just in P (Lev 11:2-28, analyzed below, as well as Lev 27:11, 27) but also in J (Gen 7:2, 8), and P's continuator H (Num 18:15). As Naphtali Meshel writes, the appearance of this opposition in the J version of the flood story implies a strong claim about the nature of the created world. The idea implicit

Lemke, and Patrick D. Miller; Garden City N.Y.: Doubleday, 1976), 414-452. In applying the lists of revealed things in wisdom literature to the category of science, the important critique of Michael Fox, "Egyptian Onomastica and Biblical Wisdom," *VT* 36 (1986): 302-310 should be borne in mind. While it hardly invalidates Stone's observations, it demands that more detailed arguments be provided for the social contexts in which the shift from ancient Near Eastern scribal "wisdom" to Hellenistic Jewish "science" took place—a demand that this essay and especially the contribution of Popović in this volume attempt to begin to answer.

[25] The Priestly theory of creation via language implicit here has never been clearly elucidated. I am preparing a study of the grammar of creation in the Priestly source, but in the meantime see the detailed discussion with bibliography in Mark S. Smith, *The Priestly Vision of Creation* (Grand Rapids, Mi.: Eerdmans, 2009).

[26] Though see the article of Meshel cited below for a demonstration that the underlying concept appears in P as well.

in the sources is that the clean-unclean differentiation that modern scholars would see as cultural is in fact based in physical nature:

> the distinction between pure and impure animal species is no innovation of Israelite religion, but was recognized in antediluvian times. Furthermore, since Noah is intuitively familiar with it, YHWH has no need to enumerate the pure and impure species. Thus it appears that the distinction does not stem from a divine decree, but rather from naturally inherent traits, easily discernible by any human being, Israelite and non-Israelite alike.[27]

Yet a contrast between nature and culture does arise within a later strand of biblical literature about the physical world. Meshel demonstrates how a nature/ culture opposition was editorially *added into* the structure of a Priestly text—in the redaction of Leviticus 11 and in opposition to the parallel set of regulations from a shared *Vorlage* in Deuteronomy 14:4-20.[28] In the Priestly portions of the text, Lev 11:1-42, 46-47[29] the editor introduced two distinct negative categories of creatures: שקץ "prohibited for consumption" and טמא "impure, ritually defiling." The distinction between the two means that not every animal that is prohibited for consumption is considered

[27] Naphtali Meshel, "Food for Thought: Systems of Categorization in Leviticus 11," *HTR* 101 (2008): 203 229 at p. 209.

[28] Meshel, "Food for Thought." On the shared *Vorlage* of Lev 11 and Deut 14, William Moran remarks that "there has long been general agreement that the *concordia discors* which careful comparison of the two passages reveals, is best explained by dependence on a common source." For earlier bibliography see his "The Literary Connection Between Lev 11:13-19 and Deut 14:12-18," *CBQ* 28 (1966): 271-277 (note that in the quoted passage Moran is referring to Lev 11:2b-23 and Deut 14:4-20, rather than the narrower portion addressed in the rest of his essay) and more recently Meir Paran, *Darkhe ha-signon ha-Kohani ba-Torah: Degamim, Shimushe Lashon, Mivnim* (Jerusalem: Magnes, 1989), 340.

[29] Lines 43-45 have long been recognized as an H insertion (see e.g. Israel Knohl, *The Sanctuary of Silence* [Minneapolis: Fortress, 1995], 69). This is clearest from the distinctively H exhortation to be holy, with an allusion to the Exodus, in line 45. Yet this division shows a striking coherence with Meshel's analysis: while terms derived from the שקץ and טמא stems are used consistently everywhere else in the chapter, here they are used indiscriminately. For literature see Meshel, "Food for Thought," 213-214 and n38.

ritually impure. For example, among the שרץ "swarming creatures" on land, which are all prohibited, only eight types are designated impure in Lev 11:29-30. Similarly, and remarkably, not every animal that is impure is prohibited for consumption! Lev 11:39-40 explain that if a permitted quadruped dies of itself, it is ritually defiling—but may be eaten, with only the routine, day-long impurity (וטמא עד הערב) which is assumed as a matter of course in daily life according to P.

Thus the old dichotomy of pure and permitted versus impure and prohibited is replaced with a four-part grid: pure and permitted, pure and prohibited, impure and permitted, impure and prohibited.[30] This new division was not practical. Because Lev 11's configuration of laws focuses on improbable situations and foodstuffs, it appears to have been conceptual, rather than normative.

This editorial rethinking of Leviticus 11 is a sign of an emergent speculative strand within Priestly thought that has begun to treat created and commanded categories separately. It stands in contrast with an older conceptual framework, shared across multiple biblical sources, that sees created and commanded as homologous, at least with respect to the categorization of animals.[31] The two different biblical treatments of created and commanded orders suggests that Second Temple Judaism inherited more than one theory dealing with the issues of *physis* and *nomos*, nature and culture. It is within the Priestly tradition that signs of a rethinking emerge, against the default biblical taxonomy of animals in which commanded is homologous with created.

How did these biblical taxonomies of the cosmos, temple and body which early Judaism inherited cohere? It has been convincingly

[30] For the argument for a further four-part distinction in Lev 11:2-8, 24-28 between species that are pure and permitted to be touched, pure and prohibited from being touched, impure and permitted to be touched, and impure and prohibited from being touched see Meshel, "Food for Thought," 216-220.

[31] For an alternative view of the Priestly relationship between creation and command, see Simeon Chavel, "'Oracular Novellae' and Biblical Historiography: Through the Lens of Law and Narrative" *Clio* 39 (2009): 12, and in greater detail "Hasifrut Hamishpaṭit Shebamiqra'" in *Sifrut Hamiqra': Mavo'ot ve-Meḥqarim* (ed. Tzipporah Talshir; Jerusalem: Yad Ben-Tzvi, 2011), 249-255.

argued that the Priestly writers saw cosmos and temple as homologous in essential ways; H emphasizes the parallel thus:

את-שבתתי תשמרו ומקדשי תיראו אני יהוה "My Sabbaths you shall observe/And my sanctuary you shall revere: I am the Lord."[32] The human body represents the third term in this homology: humans exist within and serve as well as endanger both cosmos (Gen 6:12-13) and temple (Lev 16).[33]

If the Priestly strand of the Pentateuch contains exact descriptive knowledge of the cosmos, temple, and body, they are not presented symmetrically. The first corpus is brief and unlike the latter two it is presented by an anonymous narrator, not God. This Priestly discourse begins with the narrator's treatise on the creation and structure of the cosmos—ordered through God's speech into binary divisions of things and a temporal cycle of seven (Gen 1-2:4a; again, contrast Dtr's warning against attention to and divinization of celestial phenomena in Dtr 4:19). This narrative is not spoken by God but consists chiefly of instances of God's speech narrated by an anonymous voice.

The second corpus, about the temple, has a drastically different epistemology than the first. It is framed as a divine speech that goes into extensive detail on the precise measurements and materials of the tabernacle, the ritual prototype of the temple, and its implements (Ex 25-31). Remarkably, it presents its information not as words but as a visual model (תבנית):[34]

[32] The phrase appears twice in the Holiness code, at Lev. 19:30 and 26:2. Note the striking parallels in the Priestly announcements of the completion of the cosmos and the tabernacle between Gen 1:31-2:3 and Exod. 39:32, 39:43, 40:9, and 40:33-34 analyzed by Jon D. Levenson, "The Temple and the World" *Journal of Religion* 64 (1984): 287. The crucial arguments were made by Moshe Weinfeld ("Shabbat, Miqdash, Wehamlakat H," *Beit Mikra* 21 [1977] esp. 188) and Levenson, "The Temple and the World," 286-288 and filled out with respect to Mesopotamian comparanda by Victor Hurowitz, *I Have Built You an Exalted House: Temple Building in the Bible in the Light of Mesopotamian and Northwest Semitic Writings* (Sheffield: JSOT Press, 1992). See now Richard S. Ellis, Mark J. Boda, and Jamie R Novotny, eds. *From the Foundations to the Crenellations: Essays on Temple Building in the Ancient Near East and Hebrew Bible* (Münster: Ugarit-Verlag, 2010).

[33] cf. Milgrom, "Israel's Sanctuary: The Priestly 'Picture of Dorian Gray,'" *RB* 83 (1976): 390-399

[34] Cf. the narrator's presentation of the tablernacle's construction in Ex 35-40. As we shall see below, Enoch's later perception of a *visual model* of the patterns of cosmic order in the later Qumran edition of EnAstr (4Q208, the

ככל אשר אני מראה אותך את תבנית המשכן ואת תבנית
כל-כליו וכן תעשו

"Exactly as I am **showing** you—the pattern of the Tabernacle and the pattern of all its furnishings—so shall you make it.

Note that here Moses does not "see" the תבנית on his own, but rather God causatively shows him, in the *hiphil* of the standard Biblical Hebrew verb of seeing, ראה. Moses' visions of the Tabernacle imply a different epistemology within the Priestly source. All descriptions of these visions are marked by being narrated with syntactically passive forms. Ex 25:40 reads:

וראה ועשה בתבניתם אשר-אתה מראה בהר

"See, and make, according to their pattern which you are **being shown** on the mountain," using a *hophal*, the grammatical passive of the causative.

Moses is then *shown* a rule: Ex 26:30 reads:

והקמת את-המשכן כמשפטו אשר הראית בהר

"Then set up the Tabernacle according to its rule, that you **were shown** on the mountain," also with *hophal*.

Finally, Ex 27:8 narrates Moses' vision with a morphologically active but pragmatically passive *hiphil*:

נבוב לחת תעשה אתו כאשר הראה אתך בהר כן יעשו

earliest known version c. 200 BCE, contains no narrative framework, while the edition underlying 4Q209-211, all c. 50 BCE, display evidence of a narrative framework—see further for discussion). Crucially, Enoch is *shown* the "calculations" of the luminaries' movements: ח]שבון אחרן אחזית לה "I was shown another calculation for it" in 4Q209 25 3 (corresponding to 1 En 74:1 or 78:10). And 1 En 72:1, the narrative frame of the *Astronomical Book*, describes Uriel as showing Enoch not only the physical position and temporal cycle of each luminary, but also their "names, places of origin, and months" as well as their "books."

"Make it hollow, of boards. As you **were shown**[35] on the mountain, so shall they make it."

The vast majority of the Priestly work and Deuteronomy consists of scenes where Moses *hears* God's instructions; but for Moses to *see* God's instructions requires a special sort of language. The power relationship is made evident in the text's morphology and syntax. The special grammar of seeing in the tabernacle vision has the result of denying Moses' epistemological agency: he does not even see the tabernacle under his own power, but rather is passively *shown* by God.[36]

Finally, the Priestly corpus' first set of instructions for ritual in the tablernacle ends with a manual for the observation and ritual response to discharges and affections of the human body (Lev 12-15). Unlike the tabernacle vision, the rules for discharges have a known, earthly object: the human body. They are therefore made known as verbal commands, not as a report about a passive vision of a heavenly model. Since the reader is not privileged to see the תבנית (visual model) Moses saw, both the tabernacle vision and the rules for discharges end up presented in words. The literary corpus of P transmits both of the latter to the reader entirely in divine direct discourse.

The point to take away from different framing of exact knowledge within the Priestly corpus is that there is a gap in how the three orders of knowledge are presented as being learned—the second two, about the tabernacle and the body, are framed as revealed, spoken or shown by God to be passed on by a human speaker. By contrast, the first corpus, about the cosmos, is not explicitly claimed to be revealed. This corresponds to the practical ritual function of the second two, which are represented as divinely spoken commands to be enacted word for

[35] Literally "as he showed you," but since the narrative is unambiguous that the agent of showing (God) is also the speaker (who would have to be marked in the first person), the literal translation is not plausible. This phenomenon, known as the impersonal passive, is well known in Biblical Hebrew as well as *e.g.* Medieval Latin. On both, see Paul Joüon and Takamitsu Muraoka, *A Grammar of Biblical Hebrew* (SB 27; Roma: Pontificio istituto biblico, 2006, 2nd ed.) § 128*b*.

[36] Compare the reference back to this vision in Num 8:4, "According to the vision that the LORD *showed* Moses, so he made the lampstand."

word by humans: the temple and the body are sites on which humans act ritually, while the cosmos cannot be acted upon by humans; rather, it sets the scene for all ritual. The construction of the temple and the treatment of pure and impure human bodies are chartered differently from creation.

If the opposition of "science" and "religion" is anachronistic for early Jewish texts, how can we move beyond discard the misleading old binarism? As we have seen in the case of Lev 11, we cannot assume that native categories of knowledge were uniform or stable even within a coherent corpus such as the Priestly work: there are different ways knowledge is said to be mediated, and different relationships between created and commanded orders.[37] What new analytical categories better organize the data? If we cannot find a uniform opposition in ancient Priestly (or, perhaps, Qumran) works between nature and culture, what separated their exact technical knowledge from law or ritual?

A solution to the question of what the new genres of Jewish knowledge had in common and how they patterned together may lie in attending to their status as knowledge, to precisely how they claimed to be known. In other words, to understand how ancient Jewish arts of knowledge may have been understood as sciences, it may be most helpful to focus not on an anachronistic modern concept of how scientific knowledge should be created, but on ancient concepts of how it was created. In the examples we have seen, these ancient discourses do claim that "the truth is out there" in the world, and that it becomes humanly known by observation or calculation. But as we will see, what may be most distinctively ancient and Jewish about Enochic science is its sense of non-human agency, one interestingly different from modern notions of scientific knowledge production.

[37] It is worth emphasizing that "medical" observation of the body was not separate from ritual procedures that we may consider magical. While Gen 1 and Lev 12-15 involve systematization of observed phenomena, Lev 12-15 share explicit stipulation to observe "medical" signs in the body with Num 5, a redactionally complex text which at least in its final form contains an incantation. A useful study of its editorial character is Michael Fishbane, "Accusations of Adultery: A Study of Law and Scribal Practice in Numbers 5:11-31," *HUCA* 45 (1974): 25-46.

3. Revealed Science and Evidential Grammar in Early Enochic Literature

The rise of new genres in early Judaism containing exact knowledge of the stars and body demands explanation.[38] And any historical account of the rise of these forms will need to account for their connections to old ones. Otherwise we risk turning the Hellenistic period into a black box, the distinctiveness of which is predicated on an inexplicable paradigm shift in the status of foreign knowledge and text genres.[39] We want a description fine-grained enough to account for the specificity of the forms this new knowledge takes, but broad enough to account for both their connections with and breaks from older ones.

Jonathan Ben-Dov has suggested one promising way of seeing new "scientific" genres of exact descriptive knowledge as continuous with earlier biblical ones: through the reshaping of genres. A striking feature of the oldest known Hellenistic Jewish scientific text, the *Astronomical Book* of Enoch, is its narrativization. An older, factual description of the layout and cycles of the cosmos appears to have been edited into a story involving Enoch.[40] As Ben-Dov notes, the *Astronomical Book* is not alone here, but shares the fact of narrative

[38] For an early attempt at this see Franz Cumont's 1912 *Astrology and Religion among the Greeks and Romans* (New York and London: G.P. Putnam's sons). The intellectual framework provided by Jonathan Z. Smith does not seem to have been surpassed. See his "Wisdom and Apocalyptic," in *Religious Syncretism in Antiquity. Essays in Conversation with Geo Widengren* (ed. B. A. Pearson; Missoula, MT: Scholars, 1975), 131-156, repr. in J. Z. Smith, *Map is Not Territory: Studies in the History of Religions* (SJLA 23; Leiden: Brill, 1978), 67-87.

[39] This issue is hardly restricted to the study of Judaism; in fact it is a general problem in the periodization of the human sciences, above all apparent in theories of modernity. As I summarized the arguments of the anthropologist John Kelly, "if everyone agrees that everything interesting happens in the modern period, then by definition modernity becomes very difficult to understand because it does not really come from anywhere and there is nothing to compare it to historically." Introduction, *Margins of Writing, Origins of Cultures*, 5.

[40] Ben-Dov, *Head of All Years*. For an analogous apocalyptic renarration of a preexisting descriptive genre—of myth—compare the groundbreaking analysis of the relationship between the throne-visions of Daniel, the *Book of Watchers*, and the *Book of the Giants* (demonstrating the typological priority of the editorially later vision of Daniel) by Ryan Stokes, "The Throne Visions of Daniel 7, 1 Enoch 14, and the Qumran *Book of Giants* (4Q530): An Analysis of Their Literary Relationship," *DSD* 15 (2008): 340-358.

presentation with the other two oldest known Qumran texts presenting exact knowledge, the *Book of Watchers* and the Aramaic Levi Document: "Whereas the three early scientific writings are embedded within literary frameworks and presented as part of a comprehensive patriarchal teaching, the three later items [4Q561, 4Q318, and 4Q560] lack a clear extra-scientific framework."[41] The narrative framing of exact knowledge thus did not merely legitimate a form of previously alien wisdom: it provided a new context and set of connections for its meaning and use.

The narrative framing of new knowledge in *Astronomical Enoch* is illuminating for two reasons: for the continuities it creates with earlier literature, and for the epistemological claims it makes about the knowledge it presents. The continuity it creates is clear: the angel Uriel shows Enoch the cycles of the universe, making astronomy into the content of revelation. This reframing creates a symmetry with the narrative framing of two of our three earlier corpora of exact Priestly knowledge: in the revelation of the Tabernacle's heavenly model (תבנית) and the Levitical rules for the observation and categorization of disease. But these Priestly revelations were, not accidentally, about the temple and body—not the cosmos. Here is where *Astronomical Enoch* creates something new: for the first time in a known Jewish text, astronomy is presented by a divine being, the angel Uriel. Thus the editors of *Astronomical Enoch* did not so much rupture as fill out the Priestly paradigm they inherited; P had framed exact cosmic knowledge in Genesis as spoken by a narrator, not God. *Astronomical Enoch's* presentation of exact cosmic knowledge as revealed makes it symmetrical with the earlier P presentations of revealed exact knowledge about the temple and body.

But precisely how was Enoch said to gain this revealed cosmic knowledge? A paradoxical phrase, preserved in the Aramaic of the *Astronomical Book*, allows us to be more precise about the epistemological status claimed for the astronomy revealed to Enoch.

It seems that many of the new pieces of Babylonian astronomical knowledge that Enoch learns in the *Astronomical Book* parallel the passive syntax of Moses' tabernacle vision. They are framed as חשבון

[41] Ben-Dov, "Scientific Writings in Aramaic and Hebrew at Qumran," 387.

אחזית אחרן "I was shown another calculation."[42] Grammatically a verbal phrase based on the 1cs internal passive *aphel* suffix form of חזי (most likely to be vocalized *ʾoḥzayit* or *ʾaḥzayit*),[43] it has the remarkable feature of taking the Hebrew loanword חשבון "calculation, reckoning" as an object.[44] Here we see the prime experiential categories of observation and calculation brought together in a distinctive new way. The mathematical formulae are not calculated, but the calculation (like the tabernacle's משפט) is *shown* to Enoch—an account that stands in sharp contrast with Pseudo-Eupolemus' claim that Enoch "discovered" (with *eurisko*) astrology.[45]

What is more, this phrase is not merely a part of the editorial framework of the *Astronomical Book*; it is grammatically parallel with the way new pieces of mythical cosmic geography are introduced in Enoch's otherworldly journey in the *Book of Watchers*, אחזית טור אחרן "I was shown another mountain," a phrase that appears in the singular or plural at least three times in the preserved Aramaic portions corresponding to Enoch's second otherworldly journey:

4Q204 1 xii 26-28 = 1 En 31:1-2[46]

[42] While it also appears in the *Book of Watchers* (see below), in the *Astronomical Book* the phrase is only preserved in the fragmentary passage 4Q209 25 3, corresponding to 1 En 74:1 or 78:10: ח]שבון אחרן אחזית לה די אזל "I was shown another calculation of it, when it goes ...", but it corresponds to a pattern in the Ethiopic (visible in both 74:1-2 and 78:10) in which the simple active "I saw" is used, with the angel Uriel added as agent in an explanatory phrase. Józef Milik explains that this is due to a tendency of the Greek translations, passed on to the Ethiopic, to render Aramaic passives as actives. See Milik, *The Books of Enoch: Aramaic Fragments of Qumrân Cave 4* (Oxford: Clarendon Press, 1976), 202.

[43] For a careful analysis of this grammatical form and new arguments about its vocalization see the definitive study by Edward M. Cook, "The Causative Internal Passive in Qumran Aramaic." f/c in *Aramaic Studies*. Pace Beyer, who vocalizes *'oḥzīt*, Cook points to the sporadic writing of the diphthong with 'alef in Qumran Aramaic and the presence of an uncollapsed diphthong in Biblical Aramaic. Contrast Klaus Beyer, *Die aramäischen Texte vom Toten Meer: samt den Inschriften aus Palästina, dem Testament Levis aus der Kairoer Genisa, der Fastenrolle und den alten talmudischen Zitaten* (Göttingen: Vandenhoeck & Ruprecht, 1984), s.v. חזי.

[44] Note in the Genesis Apocryphon the important role of Noah's חשבון in the completion of the first stage of his history, after ten jubilee periods, which culminates in his seeing of a vision (חזיון); see 1Q20 VI 9-11.

[45] Eusebius, *Praep. Evang.* 9:8; for a brief presentation and a comparable text from Josephus see the Appendix, below.

ואחזיא]ת̊ [טורי]ן אח̊רנין ואף בהון חזית אילנין די נ̊פ̊ק
27 [מנהון דמעא די מתקרא צרו וחלבנ]ה̊. [ול]ה̊לא מן
טוריא אלן אחזיאת טו̇ר 28 [אוחרן]

26 [... and] I [was shown] other [mountain]s, and also in them I saw trees, from which there came out 27 [sap which is called galbanu]m. [And be]yond these mountains, I was shown 28 [another] mountain ...

4Q204 1 xii 30 // 4Q206 1 xxvi 17 = 1 En 32:1

אלן כלצפון מדנחהון אחזית טורין אחרנין

... toward the northeast of these [mountains] I was shown other mountains

Most remarkably—and in contrast to the Ethiopic, which does not consistently preserve the phrase "I was shown (an)other mountain(s)", what we see here is an editorial framing device shared between the *Book of Watchers* and the *Astronomical Book*.[47] What is more, 4Q204 is the earliest clear evidence for a *collection* of books of Enoch since its fragments represent not only the *Book of Watchers* but also the *Dream Visions* and *Epistle*.[48] We find that the editors of this earliest collection of Enochic works drew on the image, and grammar, of Moses' passively gained vision (with the passive of the causative of the standard Biblical Hebrew verb of seeing) to frame Enoch's own passively gained visions (with the passive of the causative of the standard Aramaic verb of seeing).

If the use of *aphel* passives of חזי to frame revealed knowledge is both a calque from the Priestly Tabernacle vision and a distinctive editorial device shared between the *Astronomical Book* and the *Book of*

[46] The reading אחזיאת (here spelled contrastively with the *peal* active אחזית) is made plausible here by the parallel in the following sentence, as well as considerations of letter spacing in the fragments.

[47] An observation which only reinforces the conclusion of Michael Knibb that "the Greek text of Enoch, so far as it is known, and even more the Ethiopic cannot simply be regarded as translations of the original Aramaic text known from the Dead Sea fragments." See "The Book of Enoch or Books of Enoch? The Textual Evidence for I Enoch" in *The Early Enoch Literature* (ed. G. Boccaccini and J.J. Collins; Leiden: Brill, 2007), 25.

[48] 4Q204 is said to be derived from a text c. 100 BCE. Knibb, "The Book of Enoch," 26.

Watchers, then the Aramaic evidence bears on two old questions about the composition and editing of early Enochic literature. First, it means that the creators of this early literature drew more heavily on the language and imagery of the Pentateuch than has previously been acknowledged.

Categorical statements such as that of George Nickelsburg in his Enoch commentary that apart from Genesis "the rest of the Pentateuch is of little interest to the Enochic authors" will need to be revised.[49] Second, Randall Chesnutt recently reported the important discovery that Oxyrhynchus 2069, the earliest Greek manuscript of Enoch, dating from the early 4th century CE and thus at least a century older than the earliest Ethiopic version, represents a tradition in which the *Book of Watchers* was copied together with the *Astronomical Book*.[50] But the editorial pattern discussed here in the 3rd-century BCE Qumran fragments suggests that the connection between these books may well be no less than 6 centuries earlier!

Despite the significance of the discovery of the original Aramaic version of 1 Enoch at Qumran, no modern edition of the books of Enoch makes this data about the editorial framing of its visions available to the reader. Because they prefer to base their readings on the fully preserved Ethiopic manuscripts, Isaac's English translation of the Ethiopic Ms. Kebran 9, Nickelsburg and Vanderkam's English translation, and Uhlig's German translation all typically render "I saw." Disturbingly, the only edition that consistently presents the actual readings of the original sources is Milik's *editio princeps* of the Aramaic fragments.[51] This renders the fragmentary but consistent

[49] Nickelsburg, *1 Enoch 1*, 57, where he explains the exception of the historical summary in the (later) Animal Apocalypse. Nickelsburg has discussed this issue in at least four venues, including "Scripture in 1 Enoch and 1 Enoch as Scripture," in *Texts and Contexts. Biblical Texts in Their Textual and Situational Contexts. Essays in Honor of Lars Hartman* (ed. T. Fornberg and D. Hellholm; Oslo: Scandinavian University Press, 1995), 333–354; idem, "Enochic Wisdom. An Alternative to the Mosaic Torah?," in *Hesed ve-emet: Studies in Honor of Ernest S. Frerichs* (ed. J. Magness and S. Gitin; BJS 320; Atlanta: Scholars Press, 1998), 123–132; and "Enochic Wisdom and Its Relationship to the Mosaic Torah," in *The Early Enoch Literature*, 81–94.

[50] "Oxyrhynchus Papyrus 2069 and the Compositional History of 1 Enoch," *JBL* 129 (2010): 485-505.

[51] Even in this meticulous treatment, Milik's tendency to give priority to his reconstructions over the attested physical evidence occasionally obscures

evidence of the Aramaic—and thus a clue to the texts' editing, biblical referents, and epistemology—invisible.[52]

The Aramaic (as well as its antecedent Biblical Hebrew) grammar has epistemological consequences for the analysis of early Jewish views of exact knowledge. Obviously, Enoch's visions are a mode of revelation (they are involuntary, with a divine cause, using the same *aphel* first person passive of the חזי root as appears for the visionary revelation of heaven in the Aramaic Levi Document[53]). But there is a more specific epistemic value that vision has.

the epigraphically clear readings, as in his treatment of the *Astronomical Book* fragment. Because he reconstructs an unattested beginning to the sentence with the angel Uriel as the subject of an active form of חזי, he is forced to translate the clear reading of the first person *aphel* passive asyndetically as "[and Uriel demonstrated to me] a further calculation by having shown it unto me that ..." Milik, *The Books of Enoch,* 293. For a harsh verdict on Milik's practices see the statement of Greenfield and Stone that "unless the reader has recourse to the diplomatic transcription, or better yet to the plates themselves, he will not always be sure as to where restoration begins and the actual text ends; indeed he cannot always be sure as to how much there actually is in the text since Milik's readings are not always confirmed by the photographs, and by direct examination of the fragments." See "The Books of Enoch and the Traditions of Enoch" *Numen* 26 (1981): 90-91. For a further instance of problems arising from the emphasis Milik placed on reconstruction see Chesnutt, "Oxyrhynchus Papyrus 2069."

52 Isaac's translation reads this way for all of the relevant parts of the *Book of Watchers* (Chapters 31, 32: p. 28; 74: p. 53; the exception is the *Astronomical Book*'s "Then Uriel showed me" in 78:10 on p. 57). See Ephraim Isaac, "I (Ethiopic Apocalype of) Enoch" in *The Old Testament Pseudepigrapha* (ed. James Charlesworth; New York: Doubleday, 1985), vol 1. George Nickelsburg notices the Aramaic of 31:2 (see Nickelsburg, *1 Enoch 1: A Commentary on the Book of 1 Enoch, Chapters 1-36; 81-108* (Hermeneia; Minneapolis: Fortress, 2001) *ad loc* with note a, but the rest of the Aramaic manuscript evidence passes without notice. VanderKam places the Aramaic fragment at 74:1 and cites the Aramaic text in a footnote without comment but ignores it in his translation. See George Nickelsburg and James C VanderKam, *1 Enoch 2 : A Commentary on the Book of 1 Enoch, Chapters 37-82* (Hermeneia; Minneapolis: Fortress, 2012), 440. Matthew Black's edition carefully follows the Aramaic in the *Book of Watchers*, but since the *Astronomical Book* follows the Ethiopic and is relegated to an appendix the overall pattern is also invisible here. See Black, James C. VanderKam, and O. Neugebauer, *The Book of Enoch or 1 Enoch: A New English Edition with Commentary and Textual Notes.* (Leiden: Brill, 1985). The Aramaic evidence neither affects the translation nor is noted in the apparatus of S. Uhlig, *Das äthiopische Henochbuch* (Gütersloh: Gütersloher Verlagshaus, 1984)

53 For the phrase see 4Q213 2 15 (Beyer, *Aramäische Texte,* 194), corresponding to the Greek *Testament of Levi* 2:5ff.

In contrast to Biblical Hebrew, where the חזי verbal form represents a marked term for "seeing," often of dubious truth, in Standard Literary Aramaic חזי is the unmarked, default verb for "to see." Aramaic verbal phrases with חזי encode strong claims about their statements they contain: the truth of any object that this verb of seeing takes is implied to be self-evident. In linguistics this category is known as an evidential, a morphological or lexical category that connotes the speaker's assessment of the evidence for his or her statement.[54] Verbs of seeing are sensory evidentials signaling that the speaker's evidence for the truth of his or her statement is derived from the speaker's own sight.[55]

Sensory evidentials play a crucial role in the arguments that the *Book of Watchers* makes about human observation of the physical world. The preserved fragments of the book's introduction in 1 En 2-3 (4Q201 1 ii 1-3) evidence repeated play on the same verb of seeing, חזי, that structures the presentation of astronomical calculations in the *Astronomical Book* and mountains in Enoch's second otherworldly journey:

[ובמעדיהן מתחזי]ן ולא מעב[רין]בסרכן ח[זו] לארעה
וא[תבו]ננו בעבד⌉ [מן קדמיה לא]חרנה דמ[נ]ד[עם לא
ל]אשניה וכל מתח[ז]א [לכן] חזו לדגל ...

> [And] they (f) [become **seen** in their seasons], and they do not alt[er] their order. **S[ee]** the earth and consider its working! [From first to l]ast nothing changes and everything of it becomes **seen. See** the signs of...

The cosmos displays a set of unchanging cycles, conclusive visual proof of a god whose sovereignty and relationship with creation is also mediated through cycles.

[54] For the linguistic category and its application to narrative and culture see *Evidentiality: The Linguistic Coding of Epistemology* (ed. Johanna Nichols and Wallace L Chafe; Norwood, N.J.: Ablex Pub. Corp., 1986). A particularly clear set of examples is available from Quechua, a well-studied language with mandatory evidential forms spoken in Peru and Ecuador, where there are three evidential morphemes:
-mi indicating certain knowledge based on sensory experience
-shi indicating indirect knowledge based on verbal reports

What does the language of knowledge in Aramaic Enoch tell us about revelation and science in early Judaism? It is difficult if not impossible to meaningfully oppose a category of revelation to a category of science in the conceptual world of early Enochic literature. This is true not least because the very grammatical form of the texts, as well as their rhetorical style, entails that the subject of exact knowledge is both signs of divine order and something that God causes the knower to see. This editorial framing device and its evidential syntax subverts any opposition between revelation, as a mode of knowledge based on the claim "God revealed X," as opposed to science as based on the claim "I observed or calculated X."

The Aramaic phrase that introduces central units of astral knowledge in the *Astronomical Book* חשבון אחרן אחזית "I was shown another calculation" implies a category that one could call revealed science, which bridges the gap: the calculations are Aramaic-Babylonian astronomy, but the agency belongs to the angel Uriel who caused Enoch to see the calculation. אחזית טור אחרן, this phrase's editorial parallel in the *Book of Watchers*, lets us see one specific technique that the compilers of the earliest Enochic collection used to claim that mythic geography and astronomical mathematics had the same evidential status. If we understand science as knowledge gained from observation and calculation, as opposed to revelation, we see an explicit incorporation (and subversion?) of all of these modes of knowledge in the framework that unified the earliest books of Enoch.

What was lost in translation from Aramaic to Greek and Ethiopic was a significant piece of the grammar of Enoch's revealed, or

-*chi* indicating uncertain knowledge based on conjecture.
Thus example 1866 from David John Weber's *A Grammar of Huallaga (Huanuco) Quechua* (Dissertation, University of California Los Angeles, 1983) is *Qam-pis maqa-ma-shka-nki* "You hit me" would have to have either -*mi*, -*shi*, or -*chi* suffixed to the end, giving three possible meanings: with -*mi* "I saw/felt you hit me"; with -*shi* "(I was drunk at the time but) someone told me you hit me"; with -*chi* "(a bunch of people beat me up and) I think you were one of the people who hit me".

[55] Compare to other categories such as quotative evidentials, which signal that someone else is the source of the statement made. For a comprehensive, if sometimes polemical, description see Alexandra Aikhenvald, *Evidentiality* (Oxford and New York: Oxford University Press, 2004).

apocalyptic, science. Echoing the syntax of the Priestly account of Moses' tabernacle vision, this evidential phrase makes distinctive claims about human intellectual agency in what we would call scientific knowledge. Notably, these claims seem quite different from the claims made by Jewish writers in Greek such as Pseudo-Eupolemus, who attributes agency to the human mind when he says Enoch "discovered" astrology. By contrast, the earliest Enochic books claim that Enoch only learns about the stars passively, in precisely the same way Moses learns about the Tabernacle. They argue that knowledge comes from a unity of three factors: not just observation and calculation, but observation, calculation and (divine) manipulation.[56]

Conclusion

Attention to the precise Aramaic grammar of knowledge in Enochic literature, in contrast with its Hebrew Priestly forbears, gives us new evidence about the sources, composition, and epistemology of early Enochic literature. First, it corrects an earlier misunderstanding that Genesis was the main Pentateuchal source of Enoch's language and ideas: in fact, the authors of the *Astronomical Book* and the *Book of Watchers* reproduced the Priestly language of Exodus in the Tabernacle revelation, framing Enoch's cosmic revelation with a phrase that represents the precise Aramaic equivalent of the Hebrew. This may have further bearing on the currently inconclusive debates about the Mosaic, non-Mosaic, or anti-Mosaic nature of Enochic literature. Second, this shared editorial move provides evidence that the *Watchers* and the *Astronomical Book* were created with at least one of the same editorial techniques, already in the 3rd century BCE—at least 600 years before the earliest known manuscript evidence of this connection, in Oxyrhynchus 2069.

Finally, from an epistemological viewpoint: if we wish to use science as an analytical category in early Jewish thought—and I think we should—it will help to be specific about it. In what way did it count for ancient Jewish writers as authoritative knowledge of the physical world? What emerged at Qumran and later might best be understood as a "revealed science"—exact knowledge of the created world framed

[56] I thank Alexander Jones (personal communication, February 2011) for the phrase.

as divine discourse, with the role of human agency suppressed. This distinctive framing may help explain both how it emerged in a way that could claim to be continuous with earlier authoritative Jewish genres, and why this conjuncture did not produce new observations and calculations—its framing as revelation foreclosed these knowledge production mechanisms. But it would be the worst kind of anachronism to say it failed as science. Rather, it laid the foundation for a different, and quite productive, intellectual agenda.

I will conclude by returning to the provocative concept introduced by Jacob Taubes in his 1947 dissertation, *Abendländische Eschatologie*. It ties together some of the threads presented here, even as it opens up a further question. Taubes' term—"apocalyptic science"—suggests why the *Astronomical Book* could have been both the earliest known Jewish scientific work and the earliest Jewish apocalypse. Taubes was not the only one to recognize that apocalyptic calendars provided an intimate and regular connection between exact descriptive knowledge and political events—a form of knowledge that no modern definition seems to consider science. But he pointedly suggested that the real legacy of apocalyptic science may not have been in what we call science at all, but rather in a new vision of history:

> The events of the world are written on the face of the divine clock, so the point is to follow the course of world history to determine the hour of the aeon. Apocalypticism is the foundation which makes universal history possible.[57]

What if the integration of exact knowledge with historiography was as essential as mathematics or observation to the distinctiveness of early Jewish scientific thought? We would then need to consider whether Rochberg's cutting remark, that "It may only be incidental that elements with affinities to modern science are to found within the boundaries of Babylonian mathematical astronomy," may apply equally to early Jewish thought.[58]

[57] *Occidental Eschatology*, 33.

[58] Rochberg, "The Cultural Locus of Astronomy in Late Babylonia," in *Die Rolle der Astronomie in den Kulturen Mesopotamiens* (ed. H.D. Galter; Grazer

If so, the revealed science of Enoch and Qumran may have even greater continuities with Mesopotamian scholarship, one crucial task of which was to study the stars in order to know the trajectory of politics. The Assyriologist Mario Fales has recently argued that already in the Neo-Assyrian period, the practical effect of the "astronomical diaries," which correlated astronomy with a chronicle of events on earth, was to link heavenly observation "with the concept of diachrony, and more widely with the flow of political and social history."[59] In this case the later trajectory of apocalyptic and universal history may represent a return to, more than a falling away from, the ancient Near Eastern intellectual roots of early Jewish science. And the task of understanding the terms on which ancient Jewish thinkers understood the physical universe and their place in it would then beckon to us with yet more promise.

APPENDIX: Two Old Translations on Early Jewish "Science"

It is important to be aware that the importation of the category "science" into ancient Jewish texts has a long history. If the scholarly characterizations of ancient science by Alexander, Reed, and Ben-Dov are examples of bringing the category into self-conscious reflection, it was often done so unselfconsciously in earlier scholarship. This is apparent in the tendency to insert the term "science" into Hellenistic Greek accounts of what Abraham taught the Egyptians, making it easy to see him as the "first scientist" without any analogous terms appearing in the texts. The first passage where this tendency can be obseved is in Wacholder's translation of Pseudo-Eupolemus. The relevant fragment reads, in Greek and his translation,

> συζήσαντα δὲ τὸν Ἀβραὰμ ἐν Ἡλιουπόλει τοῖς Αἰγυπτίων ἱερεῦσι πολλὰ μεταδιδάξαι αὐτοὺς καὶ τὴν ἀστρολογίαν καὶ τὰ λοιπὰ τοῦτον αὐτοῖς εἰσηγήσασθαι, φάμενον Βαβυλωνίους ταῦτα καὶ αὐτὸν εὑρηκέναι, τὴν δὲ εὕρεσιν αὐτῶν εἰς Ἐνὼχ ἀναπέμπειν, καὶ τοῦτον εὑρηκέναι

Morgenländische Studien 3; Graz: GrazKult, 1993), 45.

[59] "*Maṣṣartu*: The Observation of Astronomical Phenomena in Assyria (7th Century BC)" in *The Inspiration of Astronomical Phenomena VI* (ed. Enrico Maria Corsini; ASP Conference Series 441; San Francisco: Astronomical Society of the Pacific, 2011), 370, n31.

> πρῶτον τὴν ἀστρολογίαν, οὐκ Αἰγυπτίους. (Eusebius, *Praep. Evang.* 9:8)
>
> And Abraham lived with the Egyptian priests in Heliopolis, teaching them many things. And he introduced astrology and other sciences to them, saying that the Babylonians and he himself discovered them, but he traced the discovery to Enoch. And he (Enoch) was the first to discover astrology, not the Egyptians....[60]

Wacholder's rendering of the Greek καὶ τὰ λοιπὰ τοῦτον αὐτοῖς, a phrase with the general sense of "*et cetera,*" as "and other sciences" very much begs the question of what category the things included fall into! The use of *eurisko* "find (out), discover, invent" to specify how Enoch learned astrology and other forms of knowledge is worth noting, since it attributes the agency to Enoch himself and not divine revelation—in stark contrast to what we will see in the *Astronomical Book.*[61] A similar move appears in Whiston's 1895 rendering of the parallel passage from Josephus.

> τήν τε ἀριθμητικὴν αὐτοῖς χαρίζεται καὶ τὰ περὶ ἀστρονομίαν παραδίδωσι. [168] πρὸ γὰρ τῆς Ἀβράμου παρουσίας Αἰγύπτιοι τούτων εἶχον ἀμαθῶς: ἐκ Χαλδαίων γὰρ ταῦτ' ἐφοίτησεν εἰς Αἴγυπτον, ὅθεν ἦλθε καὶ εἰς τοὺς Ἕλληνας. (Josephus, *Ant.* 1.167-168)
>
> He communicated to them arithmetic, and delivered to them the science of astronomy; for before Abram came into Egypt they were unacquainted with those parts of learning; for that science came from the Chaldeans into Egypt, and from thence to the Greeks also ...

[60] Ben Zion Wacholder, *Eupolemos. A Study of Judaeo-Greek Literature* (Cincinnati: HUC, 1974), 313–314. Similarly Gifford's 1903 translation, "and it was he who introduced astronomy and the other sciences to them."

[61] Compare similarly Josephus' representation of Abraham as *deducing* knowledge and using persuasive arguments, as analyzed by Annette Yoshiko Reed, "Abraham as Chaldean Scientist and Father of the Jews: Josephus, *Ant.* 1.154–168, and the Greco-Roman Discourse about Astronomy/Astrology," *JSJ* 35 (2004): 119-158.

Here again the word is inserted (twice!), and Feldman's more precise rendering avoids importing the category: "For these matters reached Egypt from the Chaldeans, from whence they came also to the Greeks."[62]

These examples are hardly exhaustive, but they do show how translators repeatedly—if not uniformly—caused ancient Jewish sources on astronomy and astrology to explicitly call them "science" when no parallel appears in the Greek. It may be possible to demonstrate that such a category does indeed stand behind Pseudo-Eupolemus and Josephus' accounts, but such a demonstration would need to carefully tease this category out from existing Greek terms such as *tekhne*, a type of practical skill often associated with crafts like metalworking, and existing in tension with *episteme*, pure knowledge. Indeed, the fact that astronomy and astrology are taught along with metalworking, medicine and magic by the fallen angels in 1 Enoch 7-9 would suggest that a category such as *tekhne* may have been more natural.

[62] Louis Feldman, *Judean Antiquities 1-4, vol. 3 of Flavius Josephus: Translation and Commentary*, (ed. Steve Mason; Leiden: Brill, 2000) *apud* Reed, "Abraham as Chaldean Scientist," 132-133.

5. Philological and Epistemological Remarks on Enoch's Science: Response to Papers by Seth Sanders and James VanderKam

Loren T. Stuckenbruck

The papers presented here both offer fine introductions that identify and problematize our understanding of the Enochic *Astronomical Book* as a product of emerging Jewish tradition that drew upon and departed from received traditions found in the Bible, Ancient Near East and Hellenistic ideas. To a considerable degree, these papers, despite their common focus on the *Astronomical Book* and interaction with the definitions of "science" by Philip Alexander, complement one another.

The paper by Sanders is concerned with the interplay between etic and emic notions of "science" as it attempts to draw attention to helpful heuristic categories for analysis such as "exact technical knowledge" (that embraces both the production of knowledge based on observation with the application of received knowledge to presciptive command and ritual), or Taubes' phrase "apocalyptic science", perhaps better retermed as "revealed science" or observation. Sanders argues on phenomenological grounds—and persuasively, in my opinion—that the *Astronomical Book*, in its interest in angelic agency in the transmission of calculations and information about heavenly bodies and other natural phenomena, draws on the kind of language which in the Bible priestly tradition (presented as revealed to Moses) is applied to instructions that focus on the tabnernacle/temple and the human body. The shift in the *Astronomical Book* of Enoch is, then, the use of revelation discourse in relation to cosmology. In categorizing the *Astronomical Book* as the presentation of knowledge about the cosmos by means of angelic agency, Sanders draws our attention to the passive voice of a verb (corresponding to either 74:1-2 or 78:10), the force of which has been lost through the translation process from Aramaic to Greek and into Ethiopic. He finds an analogy for this loss of the passive (revealed) nature of knowledge in the *Book of Watchers* where the shift from passive "I was shown" *'ohzayit*) to the active "I saw" (Grk. *tetheamai*, Eth. *re'iku*) can be observed (no pun intended!). This linguistic textual evidence serves Sanders to counter the third installment of Philip Alexander's definition of "science" as including "a significant element of direct observation of the physical world" which he regards as hard to reconcile with Greek "science" in which mathematics and calculations are more determinative than observation *per se*, and which he rightly thinks raises difficulties for what one means when applying the term "science" to begin with (especially when the history and philosophy of science studies are taken into account).

To Sanders' paper, I would have two comments or questions. For all the laudable interest in the paper in coming to grips with an emic, self-presentational understanding of what knowledge in *Astronomical Book* both involves and means, the term "science" continues in places to be casually employed. Given the cautions the paper enjoins upon students of antiquity with regard to using discourse about "science" at all (not least in the translations by Wacholder and Whiston, respectively of Pseudo-Eupolemos and Josephus), would one not be advised to find ways to avoid use of the term altogether? A second point I would like to raise is that, although Sanders has rightly noted the use of passive verbs for "seeing" in the *Astronomical Book* (as well as in *Book of Watchers*, etc.) as well as rendering of such verbs in the active voice by the time they are found in Ethiopic through the probably intermediary Greek, the Aramaic fragmentary texts also preserve several instances in which the Enochic seer actively "saw" this or that. What Enoch learns may ultimately be a matter of divine revelation given to him—the angel "shows" Enoch knowledge about heavenly phenomena, and, indeed, the predominant verbal form extant with the root *ḥz'* is in the passive (cf. *'oḥzayit* in 4Q209 25 3; 4Q212 1 iii 21); however, Enoch is still represented as an agent of revelation: in the narrativizing frame in which the knowledge is presented, he is the one who has shown Methuselah "everything" (*'ar'ayku-ka k^{w}ello* in 76:14; 79:1); moreover, at least in the *Book of Watchers*, the Enochic visionary can also be an active seer (4Q204 1 vi 5; 1 xii 26) "I saw trees" (where it varies with the passive "I was shown a mountain"). Thus, at least on the level of language, the shift from passive to active may already be at work in the Aramaic itself (cf. further the predominant use of the active voice in relation to Enoch's seeing in the 4Q206 and 4Q207 fragments to the *Animal Vision* as well as in the *Birth of Noah* in which Enoch is looking at the heavenly tablets).

The paper by VanderKam offers a helpful summary of what can be said about the general content of the *Astronomical Book*, based on what the extant Aramaic fragments and the Ethiopic version hold in common. He also offers a list of texts, mostly based on the Ethiopic version, in which Enoch is presented as an active seer, a revealer to Methuselah, or one to whom Uriel shows. In the presentation as a whole, VanderKam seems less concerned with definitions of "science" than Sanders. VanderKam, too, concerns himself with Philip Alexander's essay on "Enoch and the Beginnings of Jewish Interest in Natural Science". However, he does not so much question Alexander's definition of "science" itself as he evaluates several views advanced by Alexander that may be questioned: the reason for angelic agency being the domestication of alien wisdom, a purported Moses-Enoch rivalry, the use of Aramaic by priestly circles in Jerusalem, and the extent of "science" at work in the *Astronomical Book*. In countering Alexander, VanderKam's own understanding of the context within which the *Astronomical Book* arose becomes clear: (i) influenced by traditions such as Mul.Apin and Enuma Anu Enlil 14, the Enochic author is more likely to have undertaken the work "in the eastern diaspora", (ii) there is no evidence for a Moses-Enoch rivalry (something that Sanders' comparison of modes of knowledge in the priestly tradition in the Pentateuch with *Astronomical Book* would seem also to bear out); (iii) the use of an angelic mediator for the astronomical material is "exegetical" (Enoch walked with "elohim") rather than an attempt to give alien wisdom a place within Jewish tradition; and, to quote VanderKam, (iv) "The science in the Astronomical Book of Enoch meets the criteria that Alexander lists but does so more fully than he indicates." Although with Alexander, VanderKam acknowledges that *Astronomical Book* has probably reduced complex phenomena regarding the moon and sun into overly simple patterns, he still claims that "the science contained in it [i.e. the book] is based ... on observation".

In this last point, we may notice where VanderKam and Sanders part company. Sanders does not do away with the category of "observation" altogether, but emphasizes this as "revealed". VanderKam's emphasis seems to move in the other direction. Though arguing a certain indebtedness on the part of the writer to Genesis 1 and 5 and the Flood narrative as well as to the approach to the fixity of the created order found in Jeremiah 5:22-23 and 31:35-36, VanderKam places more emphasis on the writer's claim that the knowledge of the book is based on "observation". As such, the *Astronomical Book* belongs to what Drawnel refers to as "didactic literature" and its purpose, though presented simply as knowledge transmission by a father to a son or teacher to pupil, would have been "to produce something very schematic to provide basic knowledge to aspiring priests or others whose work would require some knowledge about the structure of the creation." Here VanderKam implies something that Sanders' paper does not: knowledge for priests "whose work would require" it suggests that for VanderKam the 364-day scheme was, at least in the eastern diaspora a practical (not simply an ideal) calendar. One might be interested in whether Sanders concurs with this, especially since Sanders finds in the *Astronomical Book* the bringing together of aspects from different spheres of "exact descriptive knowledge" (cosmos, temple and the human body), thus relativizing the distinction between "observation" of nature and prescriptive command/ritual with which language of "revealed" is more immediately associated.

Within the discussion about the degree to which the Enochic *Astronomical Book* relates to ancient "science", a significant framework —a framework that is specifically Jewish and not "science" in any way, not even in the ancient world—should not be forgotten. The book opens at 1 Enoch 72:1 with the following superscript: "The book about the motion of the heavenly luminaries all as they are in their kinds, their jurisdiction, their time, their name, their origins, and their months which Uriel, the holy angel who was with me (and) who is their leader, showed me. The entire book about them, as it is, he showed me and how every year of the world will be forever, *until a new creation* lasting forever is made."[1] The italicized words indicate that, whatever the value of revealed knowledge about the heavens may consist of, that knowledge is considered to be a *temporary* structure that at a decisive future time will be replaced by a more eternal one. Thus in the *Astronomical Book* we have to do with revealed knowledge (according to what is disclosed to the Enochic seer through the agency of Uriel) and a future, more eternal order of things designated as a "new creation". Revealed knowledge, then, is not necessarily permanent; what is revealed is, as a matter of principle, provisionary. We may legitimately ask, then, how much this opening of the book remains operative in everything that follows and, further, implies an eschatological framework that validated discourse that otherwise may have shown heavy influence from existing traditions of observing heavenly bodies that circulated in the Ancient Near East and to which the Enochic author(s) here fell heir.

I have learned a great deal from both these papers. They both argue that the notion of "observation" is operative in the claims being made in the *Astronomical Book*. However, our papers seem to differ with regard to the kind of observation that characterizes the way the book presents itself and they seem to differ regarding the origin of angelic mediation that the book sustains.

[1] The translation is that of George W.E. Nickelsburg and James C. VanderKam, *1 Enoch: A New Translation* (Minneapolis: Fortress, 2004), 96.

6. Ideals of Science: The Infrastructure of Scientific Activity in Apocalyptic Literature and in the Yahad

Jonathan Ben-Dov

1. Introduction[1]

The fragments of scientific literature discovered among the Dead Sea Scrolls justify the title of the present book as a field of study. While previous studies in the field of the history of science pointed to systematic scientific activity among Jews only as early as the Islamic period,[2] it is now clear that science existed earlier in Jewish history. After the recent publication and discussion of the scientific material in the Dead Sea Scrolls, it is now time to access this corpus by means of more general questions: why is it that scientific activity existed in the Dead Sea Scrolls community? What was the epistemological infrastructure that prompted scientific creativity in this milieu? What were the sources for the science in the scrolls, and how were these sources carried further? A wider question is whether science was equally integrated in other branches of early Judaism, whose writings were not lucky enough to be deposited in the caves. In other words, what is it in the Qumran Yahad that brought about a creative scientific vocation?

What are the necessary conditions for the establishment of a scientific culture? One may seek to answer this question by means of sociological tools, tracing the social and political framework that gives power, prestige and funding to an institutionalized scientific organ

[1] Work for the present article has been supported by the Israel Science Foundation, grant number 527/08. I am very much indebted to Seth Sanders for his illuminating remarks on earlier drafts of this paper.

[2] Y. Tzvi Langermann, "On the Beginnings of Hebrew Scientific Literature and on Studying History through 'maqbilot' (parallels)", *Aleph* 2 (2002): 169-189. The present article does not deal with the use of the term 'science' with regard to the ancient Jewish material. The philosophical justifications for this use are discussed at length in the introduction to the present volume.

within society;[3] or by means of historiography, as many recent histories of science do.[4] Although one may venture to determine the role of various practitioners of knowledge in ancient societies,[5] I believe that the main efforts should rather be directed towards the ideological aspect: what is it that prompts pre-modern writers to study and promote scientific knowledge about the world? How do they conceive of 'the World' and of the ways assigned to a human being to explore it? What are the sources of knowledge? Whence the epistemic authority of the human prerogative to study the world?[6] What are the historical or mythological precedents for this kind of vocation?

Significant motivation for the present paper—as well as for the initiation of this conference—arose from some recent historiographical work, which opened the way for a great upsurge in the study of pre-modern science among Jews. This work was primarily done by Amos Funkenstein and later David Ruderman,[7] whose categories of thought in many ways set the stage for the present discussion. From Funkenstein I learned that forms of science exist also outside the epistemological cosmos of the present-day, western and secular science, and moreover, that the roots of modern science lie

[3] For example Jürgen Habermas, "Science and Technology as Ideology," *Towards a Rational Society* (trans. Jeremy J. Shapiro; Boston: Beacon, 1971); Pierre Bordieu, "The Forms of Capital," in *Readings in Economic Sociology* (ed. N.W. Biggart; Oxford: Blackwell, 2008), 280-291.

[4] Simon Schaffer and Steven Shapin, *Leviathan and the Air-Pump: Hobbes, Boyle, and the Experimental Life* (Princeton: Princeton University Press, 1985).

[5] For example, using variegated methodologies: Anthony J. Saldarini, *Pharisees, Scribes and Sadducees in Palestinian Society: A Sociological Approach* (Edinbourgh: T & T Clark, 1985); Lester L. Grabbe, *Priests, Prophets, Diviners, Sages: A Socio-historical Study of Religious Specialists in Ancient Israel* (Valley Forge, Pa.: Trinity Press International, 1995); Carol A. Newsom, *The Self as Symbolic Space: Constructing Identity and Community at Qumran* (STDJ 52; Leiden: Brill 2004).

[6] For the term 'epistemic authority' see Arnon Keren, "Epistemic Authority, Testimony and the Transmission of Knowledge," *Episteme: A Journal of Social Epistemology* 4 (2008): 368-381.

[7] Amos Funkenstein, *Theology and the Scientific Imagination from the Middle Ages to the Seventeenth Century* (Princeton: Princeton University Press 1986); David B. Ruderman, *Jewish Thought and Scientific Discovery in Early Modern Europe* (New Haven and London: Yale University Press 1995). As it turns out, I was unaware that my motivation echoes Funkenstein's until notified by Micha Perry of Yale University. I am deeply grateful to him for an illuminating remark.

with the theological imagination no less than they lie with the rise of empiricism. The tools developed by Ruderman and others to study the scientific imagination in the medieval and early modern periods can now be applied—with some modifications—to the materialization of scientific thought in early Judaism.

Do the Dead Sea Scrolls attest to a creative scientific culture? Take for example the vast range of rabbinic literature, where one finds a huge variety of what we would call scientific knowledge: cosmography, mathematics, geography, astronomy, biology, medicine, etc.[8] Yet there remain some characteristics in the Dead Sea Scrolls which distinguish them from the knowledge collected in the Talmud. A central characteristic is the use of scientific genres, i.e. complete treatises dedicated to systematic scientific knowledge, as opposed to sporadic statements embedded in other genres. While it is still often claimed that the first systematic Jewish science books were such late-antique to early-medieval treatises as *Sefer Yetzirah*, *Mishnat ha-Middot*, *Midrash Konen*, *Baraita de-Shmuel*, *Sefer Asaf haRofe*, etc., we are now aware of some previously unknown—or at least not enough appreciated—Jewish scientific texts: the *Book of the Luminaries* in 1 Enoch, the corpus of calendars and *mishmarot*, and some texts which combine astrology and physiognomy.[9] Another central characteristic of scientific culture in the scrolls is creativity: what share of the

[8] Much of this material was collected in the late 19th or early 20th century by scholars who sought to demonstrate the Jewish mastery of science in ancient times, at least partly for apologetic reasons. The works thus vary significantly in their accuracy and reliability: e.g. Julius Preuss, *Biblisch-talmudische Medizin: Beiträge zur Geschichte der Heilkunde und der Kultur überhaupt* (Berlin: Karger, 1923); William M. Feldman, *Rabbinical Mathematics and Astronomy* (New-York: Hermon Press, 1978). More recently see the articles by Kottek, Safrai and Ophir-Shemesh in S. Safrai et al., *The Literature of the Sages* (CRINT 3/2; Assen: Van Gorcum 2006), 485-520. Reuven Kipperwasser, "Body of the Whore, Body of the Story and Metaphor of the Body" in *Introduction to Seder Qodashim: A Feminist Companion on the Babylonian Talmud* ed. Tal Ilan et al. (A Feminist Commentary on the Babylonian Talmud; Tübingen: Mohr Siebeck, 2012), 305-319.

[9] One may possibly add medicine to the scientific skills of the Dead Sea Scrolls community. See recently Joan E. Taylor, "Roots, Remedies and Properties of Stones: the Essenes, Qumran and Dead Sea Pharmacology," *JJS* 60 (2009): 226-244; Ida Fröhlich, "Medicine and Magic in Genesis Apocryphon. Ideas on Human Conception and its Hindrances," *RQ* 25 (2011): 177-198.

knowledge presented in the literary corpus continues the preceding scientific tradition, and what share proceeds from it to introduce some novel concepts. The novelty sought here is not what we call 'progress', i.e. irreversible advancement towards an absolute scientific truth. Rather, we see examples in which the scientific knowledge is so deeply integrated into the ideological and literary texture of the Dead Sea Scrolls community that it produced new paths in astrology, for example. These paths constitute unique attestations of applied science for the use of the Yahad, and attest to the extent of acceptance of the scientific discourse in this environment.

The above-noted features support the distinction between the science in the scrolls and that in later, Talmudic or medieval Jewish literature. While the latter science is not essentially Jewish, but rather "A Science practiced by Jews", the scientific interests in the scrolls are profoundly Jewish and apocalyptic in their outlook, as they werei fashioned by Jewish-sectarian scholars for religious and cultic use, and were meant to answer the particular needs of a concrete Jewish community.

The main aim of the paper, therefore, is to describe and account for the intellectual climate that prompted the formation of a creative scientific environment in the Dead Sea Scrolls community. The community will be designated here by the neutral term *Yahad*, used by community members to refer to themselves.[10] The Yahad is a specific

[10] The use of this term, as well as any talk on the community that produced the Dead Sea Scrolls, requires a clarification with regard to the basic questions of the identity of the Yahad: does it correspond to the Essenes mentioned in Greek sources? To the community of the *Damascus Covenant*? Did it reside in the site of Qumran, adjacent to the caves where the scrolls were found? However, the scope of the present discussion does not suffice to give full answer to these questions. I generally subscribe to the Essene theory, as recently modified by John Collins and Alison Schofield: Alison Schofield, *From Qumran to the Yahad: A New Paradigm of Textual Development for the Community Rule* (Leiden : Brill, 2009); John J. Collins, *Beyond the Qumran Community: The Sectarian Movement of the Dead Sea Scrolls* (Grand Rapids, Mi.: Eerdmans, 2010). The Yahad is an umbrella term designating various small groups across Palestine of the 1st century BCE-1 century CE, with one representative, possibly an elite group of the Yahad, residing at Qumran. This group produced or at least collected most of the scrolls represented in the caves. It essentially overlaps the group described in the *Damascus Document*. The entire discussion is necessarily quite limited since most probably the various sects were not clearly departmentalized as we modern

cultural and societal entity, quite limited in scope. It is thus crucial to trace the antecedents of the Yahad's scientific outlook within larger movements of Judaism of the Hellenistic period: Wisdom and Apocalypticism, or for the present purposes: The Enochic literature, the apocalyptic-sapiential Aramaic texts from cave 4, the Hebrew wisdom text 4QInstruction (*musar la-mevin*), and to a certain extent also the Wisdom of Ben Sira. While these streams of tradition triggered the commitment of the Yahad to science, scholarship should also account for the transformation of this tradition in the specifically sectarian thought and practice.[11]

2. Scientific Texts and Themes in the Scrolls

In one of Flavius Josephus' reports on the Essenes, he writes (War 2.136):[12]

> They apply themselves with extraordinary zeal to the study of the ancients choosing, above all, those which tend to be useful to body and soul. In them they study the healing of diseases, the roots offering protection and the properties of stones.

Based on this paragraph, and others similar to it, the Essenes gained the reputation of a group which excelled in divinatory and para-scientific activity. Thus for example in an intriguing remark made by William Albright in 1940: [13]

scholars expect them to be; see Eyal Regev, *Sectarianism in Qumran: a Cross-Cultural Perspective* (Berlin: de Gruyter, 2007).

[11] On the molding of earlier traditions in the framework of the Yahad see recently John Collins, "Tradition and innovation in the Dead Sea Scrolls," in *The Dead Sea Scrolls. Transmission of Traditions and Production of Texts* (eds. S. Metso, H. Najman and E. Schuller; STDJ 92; Leiden: Brill, 2010), 1-23.

[12] Translated by Geza Vermes and Martin Goodman, *The Essenes According to the Classical Sources* (Sheffield: JSOT Press, 1989), 43. Cf. Ronald Bergmeier, *Die Essener-Berichte des Flavius Josephus. Quellenstudien zu den Essenertexten im Werk des Jüedischen Historiographen* (Kampen: Kok Pharos, 1993), 87, 96-97.

[13] William F. Albright, *From the Stone Age to Christianity* (Baltimore: Johns Hopkins University Press, 1940); Stephen J. Pfann, "The Writings in Esoteric Script from Qumran" in *The Dead Sea Scrolls Fifty Years after their Discovery* (eds. L.H. Schiffman and E. Tov; Jerusalem: IES and The Shrine of the Book, 2000), 177-190, 289. See also Jonathan Ben-Dov, *Head of All Years: Astronomy and Calendars at Qumran in their Ancient Context* (STDJ 78; Leiden: Brill, 2008),

> They further possessed an extensive esoteric literature, access to which was only allowed members of the order. According to Josephus, they were interested in the virtue of plants and stones, they possessed an elaborate angelography... and they attached great importance to the art of predicting the future, in which they seldom made mistakes... we can infer that the Essenes, in opposition to virtually all pre-cabalistic Jews, were believers in astrology, which harmonizes just as well with their strict pre-destinarinism... It seems probable that the Essenes represent a sectarian group which had migrated from Mesopotamia to Palestine after the victory of the Maccabees. This theory would explain their interest in the virtues of plants and stones..., their attention to divination and astrology... since all of these points were characteristic of Mesopotamian practice.

Albright's statement is based solely on the report by Josephus (in fact it somewhat exceeds it as Josephus is never explicit about astrology and divination). Today we are in the position to assess its validity with regard to the actual collection of scrolls from Qumran. Indeed one finds in the scrolls a pronounced interest in the sciences, as attested in both scientific texts proper and in the frequent mention of scientific themes. While some of this material is in Aramaic and seems to have preceded the Yahad, other parts of it are in Hebrew and are more clearly sectarian. In a previous article I summarized the evidence for scientific activity in the Aramaic texts, which includes the following:[14] astronomy, astrology, physiognomy, geography, metrology. Hebrew scrolls from Qumran carry further the interest in astronomy and astrology, which will be discussed here at some length.

255.

[14] Jonathan Ben-Dov, "Scientific Writings in Aramaic and Hebrew at Qumran: Translation and Concealment" in *Aramaica Qumranica: Proceedings of the Conference on the Aramaic Texts from Qumran in Aix-en-Provence 30 June - 2 July* 2008 (eds. K. Bertholet and D. Stökl Ben Ezra; STDJ 94; Leiden: Brill, 2010), 379-399, 381-384 with full bibliography.

Astronomy

Astronomy was woven into the sectarian texture, primarily in the calendrical texts.[15] A notable feature of these texts is that they not only give the practical aspects of calendar making, those that are required for a quotidian routine, but rather transfer the calendrical discourse to a more elevated level, both in terms of ritual status and in terms of the scientific discourse. The meticulous anchoring of the calendar in the service cycle of the priestly families (*mishmarot*) supplies the ritual context. In addition, central calendar texts from Qumran are framed by statements on the creation of the world and the place of the luminaries at that time (4Q319 IV 10-11; 4Q320 1 i 1-5; 4Q320 3 i 10). The calendars contain detailed rosters of lunar phenomena, which cannot be explained as part of a normative calendar but should rather be seen as an astronomical apparatus.[16] It was important for calendar experts of the Yahad to include astronomical calculations—mostly very schematic—in their agenda, back to back with ritual concepts like priests and festivals.

Astrology and Physiognomy

Astrology is often mentioned or implied in writings of the Yahad. Except for the systematic presentation of astrological teachings in texts like 4Q186 (on which see below), there are recurrent references to astrological themes, as astrology acquired a central place in the predestination doctrine of the Yahad. This seems to be the case already in the wisdom texts from Qumran, especially 4QInstruction and the 'Book of Mysteries', texts which probably preceded the Yahad and were strongly embraced in it.[17]

[15] For an edition of the calendrical texts from Qumran see Shemaryahu Talmon, Jonathan Ben-Dov, and Uwe Glessmer, *Qumran Cave 4 XVI. Calendrical Texts* (DJD XXI; Oxford, Clarendon Press, 2001).

[16] See Jonathan Ben-Dov, "Lunar Calendars at Qumran? A Comparative and Ideological Study" in *Living the Lunar Calendar* (eds. J. Ben-Dov, W. Horowitz and J. Steele; Oxford: Oxbow, 2012), 173-189. For the astronomical value of these lists see Jonathan Ben-Dov and Wayne Horowitz, "The Babylonian Lunar Three in Calendrical Scrolls from Qumran," *ZA* 95 (2005): 104-120; Ben-Dov, *Head of all Years*, 197-243.

[17] For 4QInstruction see Eibert J.C Tigchelaar, *To Increase Learning for the Understanding Ones : Reading and Reconstructing the Fragmentary Early Jewish Sapiential Text 4QInstruction* (STDJ 44; Leiden: Brill, 2001), 247-248; Matthew J. Goff, *Discerning Wisdom: The Sapiential Literature of the Dead Sea* Scrolls (VTSup

In the wisdom texts one finds statements on the general evaluation of a person's role or destiny, or of the world's destiny, as this role corresponds to a preexistent cosmic order.[18] This preexisting order, or, as Elgvin calls it "a comprehensive word for God's mysterious plan for creation and history" is denoted by the enigmatic and polyvalent term רז נהיה, 'the mystery that becomes'.[19] However, the enigmatic term and the various contexts for its use do not disclose much on the nature of the mystery, or on the questions how it is to be fathomed by a human being. Usually the commandment is to deduce the *raz* by observing the wondrous deeds of god, either in history, in nature or in the fate of individuals, but no specific mechanism is indicated.[20] The terminology of 'the Book of Mysteries' and

116; Leiden: Brill, 2007), 61-65; Jean-Sébastian Rey, *4QInstruction: sagesse et eschatologie* (STDJ 81; Leiden: Brill, 2009), 335. In contrast, Menahem Kister, "Wisdom Literature and its Relation to Other Genres," in *Sapiential Perspectives: Wisdom Literature in Light of the Dead Sea Scrolls: Proceedings of the Sixth International Symposium of the Orion Center for the Study of the Dead Sea Scrolls and Associated Literature, 20-22 May, 2001* (eds. J.J. Collins, G.E. Sterling and R. Clements; STDJ 51; Leiden: Brill 2004), 13-47 is inclined to view 4QInstruction as closer to the writings of the Qumran community, and so is Bilhah Nitzan, "Key Terms in 4QInstruction: Implications for Its Ideological Unity," *Meghillot* 3 (2005): 120-121 (in Hebrew).

[18] Nitzan, "Key Terms in 4QInstruction," 105-106 points out the difference between different compositions in this respect: while the Book of Mysteries (1Q27, 4Q299, 4Q300) discusses the fate of all nations, the contexts in 4QInstruction are oriented towards individual fate.

[19] Torleif Elgvin, quoted by Kister, "Wisdom Literature," 31. In terms of verbal tenses, the participle of *hyh* in this phrase connotes the aspect of an unfinished act: the mystery that becomes. See the cogent analysis in Matthew J. Goff, *The Worldly and Heavenly Wisdom of 4QInstruction* (STDJ 50; Leiden: Brill 2003), 51-79 with earlier literature adduced there. Goff chose the translation "the mystery that is to be," and Kister: "secret (or: mystery) of (things) to come" (Kister, "Wisdom Literature," 31), based on similar formulations in Ben Sira 42:19, 48:25. In contrast, Rey, *4QInstruction*, 284-292 stresses the ambiguity of the participle *niphal*, and demonstrates how both past and future aspects are produced by this participle in various contexts. In order to bypass this ambiguity he prefers a nominal form "le mystere de l'existence," despite the fact that the rigidity of the nominal form misses the dynamics of the participle (Rey, *4QInstruction*, 292 n.44). For the esoteric message of the word רז see Samuel I. Thomas, *The "Mysteries" of Qumran: Mystery, Secrecy, and Esotericism in the Dead Sea Scrolls* (Atlanta: SBL, 2009).

[20] For a deduction of the *raz* from the observation of history and nature see 4Q471 i 1-13 (with parallels in 4Q418 and 4Q418a; see Tigchelaar, *To Increase Learning*, 52; and more generally Kister, "Wisdom Literature," 29-35). Lange claimed that the *raz* is to be found in writing, within the mysterious ספר

4QInstruction suggests that—at least in some cases—this observation should be achieved by means of astrology. Strong indications come forth from the terms מולד and בית מולדים, which appear in both wisdom texts, often closely associated with רז נהיה. Despite its simple meaning 'birth', the context often requires that these terms convey the more technical meaning of 'nativity' or even 'horoscope', as recorded in some later Aramaic literature.[21] In these cases, as noted quite clearly by Kister and Baumgarten, the simple meaning 'birth' is insufficient to carry the burden loaded upon it in such polyvalent contexts as the following: [22]

מ]חשבת בית מולדים פתח לפ[ניהם/נינו

(God) has opened before [us / them] the art of nativity[23]

ההגו and ספר זכרון, basing himself on the statements in 4Q417 1 i 14-18 and on reverberations of the verb חרת: see Armin Lange, *Weisheit und Prädestination: weisheitliche Urordnung und Prädestination in den Textfunden von Qumran* (STDJ 18; Leiden: Brill, 1995), 19. Lange sees the entire section in 4Q417 1 i as a continuous statement on the nature of the *raz*. However, the section mentioning the written books (4Q417 1 i 13-18) is clearly separated from the neighboring sections by the headings ואתה (בן) מבין (lines 13 and 18), as well as by a *vacat* in line 18. This section does *not* mention רז נהיה at all, and is thus not a direct sequence to the previous section.

[21]For מולד as an astrological technical term see Matthew Morgenstern, "The Meaning of בית מולדים in the Qumran Wisdom Texts," *JJS* 51 (2000): 141-144. It should be noted, however, with Tigchelaar, *To Increase Learning*, 238, that not every mention of מולד in Yahad literature should be automatically associated with astrology, since some contexts clearly defy this meaning. The following discussion will dwell on several such contexts.

It should also be noted that it is not clear whether the cognate terms in Syriac and Mandaic quoted by Morgenstern reflect the exact meaning of Greek *horoskopos* (see below). Rather, I suspect that they reflect the less-specific meaning 'nativity', which is astrological nonetheless but not rigorously technical. I thank Alexander Jones for pointing out this matter to me.

[22] Joseph M. Baumgarten, "Qumranic and Astrological Terminology in Musar leMevin," *Tarbiz* 72 (2003): 321-328 (in Hebrew); Kister, "Wisdom Literature," Cp. Goff, *Discerning Wisdom*, 24, 81.

[23] Kister, "Wisdom Literature," 28 connects this passage with the debate on the origins of astronomy and the divinatory arts - between good wisdom and illicit knowledge. See also below. Kister reads לפ[נינו while Schiffman in DJD 20: 41-42 reconstructs לפ[ניהם; see Kister, "Wisdom Literature," 28 n. 71.

הבט ברז] נהיה וקח מולדי ׄישע ודע מי נוחל כבוד ועׄ[ו]ל

> [Gaze upon the mystery] that becomes, and comprehend the nativities of salvation, and know who is to inherit glory and (and who-) e[vi]l. (4Q417 2 i 10-11)[24]

וברז [נ]הׄיׄהׄ דרוש מולדיו וׄאׄז תדע נחלתו

> And by the mystery that becomes study the nativity thereof and then you shall know its inheritance (=fate) (4Q416 2 iii 9-10)

The addressee is commanded to enquire the רז נהיה by means of investigating various nativities: either with regard to the general history of the world or with regard to the fate (lit. inheritance, equivalent to Greek *κλῆρος, κληρονομία*) of a specific individual.[25] Baumgarten suggested that a similar command is issued to a person to calculate his wife's nativity (4Q416 2 iii 20-21), and possibly also that of his newborn child (4Q415 11 11). The passage about the wife reads:[26]

אשה לקחתה בריׄשכה קח מולדיׄ]ה ..... [מרז נהיה

> You have taken a woman in your poverty, inquire [her] nativities [....] from the *raz nieyeh*

Kister highlights the use of מולדים in 4Q299 frg 5, which in his opinion has "explicit astrological overtones" because it is appears in conjunction with theological-cosmological poetry about the regular alteration of light and darkness.[27] In some way therefore, astrological

[24] Hebrew transcriptions of 4QInstruction follow Tigchelaar, *To Increase Learning*, 47, 55. The reconstructed portions are practically certain based on parallels in other copies of 4QInstruction. English translations are mine.

[25] See TDNT s.v. *κλῆρος, κληρονομία*, III: 760-764, 779-785; further Nitzan, "Key Terms in 4QInstruction," 109-120. Nitzan, however, does not dwell on the astrological aspects of this word. I believe that the growing frequency of this word in wisdom writings from the Second Temple period (Tigchelaar, *To Increase Learning*, 239) is connected with the growing awareness to questions of pre-destination and free will.

[26] Baumgarten, "Qumranic and Astrological Terminology," 324.

[27] Kister, "Wisdom Literature," 45. Kister further connects this passage with other apocalyptic passages in which a cosmological section introduces an eschatological admonition: 1 Enoch 2-5 as well as the introduction to

knowledge is part and parcel of revealed wisdom in these texts, and is recurrently associated with the רז נהיה.

It is remarkable that the texts never indicate the method invoked in order to study the nativity, but rather limit themselves to general statements on astrological rulings. This is even more disturbing due to the fact that no 'horoscope' proper was found at Qumran. Even in 4Q186, the closest exemplar of a specific astrological technique, one searches in vain for what historians of astrology would call a horoscope.[28] Not only the technique, but also the role of astrology remains implicit or even veiled in the overall formulation of the texts. The fact that the astral mechanisms are seldom explicit in the scrolls is due to the fact that the judgment was in the hands of the *Maskil* alone (see quotes from 1QS below). Yet it remains clear, in my eyes, that astrology reflects the true preordained order of the world. 4QInstruction often advises the addressee how to cope with his preordained fate in an active yet legitimate way. Since in this text the fate is determined mainly—or at least to a great extent—by astrology, one is justified to say that the move in 4QInstruction resembles that of some later stoic literature, which advocates a similar move.[29]

4QInstruction, as reconstructed by Tigchelaar, *To Increase Learning*, 175-193, and discussed further by Rey, *4QInstruction*, 228-276. For the role of Laws of Nature in apocalyptic admonitions see Lars Hartman, *Asking for a Meaning. A Study of 1 Enoch 1-5* (CB.NT 12; Lund: Gleerup, 1979); Michael E. Stone, "The Parabolic Use of Natural Order in Judaism of the Second Temple Age," in *Gilgul. Festschrift R.J.Z. Werblowsky* (Numen Supplement 50; eds. S. Shaked, D. Shulman and G.G. Stroumsa; Leiden: Brill, 1987), 298-308.

[28] On horoscopes see Otto Neugebauer, and Henry B. van Hoesen, *Greek Horoscopes* (MAPS; Philadelphia: American Philosophical Society, 1959); Alexander Jones, *Astronomical Papyri from Oxyrhynchus: (P. Oxy. 4133-4300a)* (Philadelphia: American Philosophical Society, 1999), 10-11; Matthias Albani, "Horoscopes in the Qumran Scrolls," in *The Dead Sea Scrolls after Fifty Years: a Comprehensive Assessment* (eds. J.C. VanderKam and P.W. Flint; Leiden: Brill, 1999), 280-282. The so-called 'Babylonian Horoscopes' in Francesca Rochberg, *Babylonian Horoscopes* (TAPS 88; Philadelphia: American Philosophical Society, 1998) do not correspond with this form either, but they are closer to the Greek format in that they indicate a full set of stellar and planetary positions at a given time for one specific person. It is important to note that not one planet or star is mentioned by name in the DSS outside the zodiac names in 4Q318 and 4Q186.

[29] For the stoic encounter with astrology see the old but still valuable survey by Franz V.M. Cumont, *Astrology and Religion among the Greeks and Romans* (New York: GP Putnam, 1912). For a preliminary comparison of the determinism in Qumran wisdom with Stoicism see Martin Hengel, "Qumran

Central texts of the Yahad continue the sapiential-deterministic line of 4QInstruction, while concealing the explicit role of astrology. Thus in the Discourse on the Two Spirits (1QS III 13 – IV 26) the metaphysical order of the world is entirely administered by angels and demons rather than by stars. However, hints for astrological determination of human fate remain in the introduction to the Discourse (1QS III 13-15), stating the roles of the *Maskil*:

למשכיל להבין וללמד את כול בני אור בתולדות כול בני
איש לכול מיני רוחותם באותותם למעשיהם בדורותם

> To the *Maskil*. To instruct and teach all the Sons of Light concerning the nature of all the sons of man, according to all the kinds of spirits revealed in the character of their deeds during their generations[30]

This passage refers to the תולדות of all humanity, a term based on the root ילד which in later Jewish literature is associated with the deterministic rulings of physiognomy. It may be connected with Greek φύσις and Latin *natura*, terms which relate to the fixing of human character at the time of birth.[31] In addition, later in the same sentence from 1QS human beings are classified by means of their אות, 'sign', a term which carries astral connotations. As Baumgarten noted in the above mentioned article, as well as Mladen Popović and Philip Alexander, astrological categories played a role in sorting out the

and Hellenism," in *Religion in the Dead Sea Scrolls* (eds. J.J. Collins and R. Kugler; Grand Rapids, Mi.: Eerdmans, 2000), 46-56; Corrado Martone, "Qumran and Stoicism: An Analysis of some Common Traits," in *The Dead Sea Scrolls Fifty Years after their Discovery 1947-1997* (eds. L.H. Schiffman, E. Tov and J.C. VanderKam; Jerusalem: IES and the Shrine of the Book, 2000), 617-622; further David Flusser, *Judaism of the Second Temple Period. Volume 1: Qumran and Apocalypticism* (Grand Rapids, Mi.; Jerusalem; Eerdmans: Hebrew University of Jerusalem / Jerusalem Perspective, 2007), 114-139. I consider some parallel notions between Stoicism and Qumran sectarian thought to be a fruitful field for future study.

[30] Translation follows partly that of Elisha Qimron and James H. Charlesworth, "Rule of the Community (1QS)", in *Rule of the Community and Related Documents* (ed. J.H. Charlesworth; SSHAGT 1; Tübingen / Louisville: Mohr Siebeck / John Knox, 1994), 15. The latter part follows the translation by Knibb, quoted ibid. note 61.

[31] See detailed bibliography in Mladen Popović, *Reading the Human Body: Physiognomics and Astrology in the Dead Sea Scrolls and Hellenistic-Early Roman Period Judaism* (STDJ 67; Leiden: Brill, 2007), 180 n. 29.

candidates for membership in the Yahad and even members of the Yahad themselves.[32] Thus we read elsewhere about the role of the *Maskil* or the *Mebaqqer* in judging and assigning a place to community members (CD 13:11-12):

וכל הנוסף לעדתו יפקדהו למעשיו ושכלו (following 4QD[a])
וכוחו וגבורתו והונו וכתבוהו במקומו כפי נחלתו בגורל
הא]ור

> And whoever joins his congregation, let him examine him with regard to his works and his intelligence, his strength and might, and his wealth. Let them inscribe him in his place according to his inheritance in the lot of lig[ht][33]

The classification of Yahad members was done according to their relative share in the assigned portions of light. This procedure is reflected in the astrological-physiognomic teaching of the scroll 4Q186, on which see below. A scientific mechanism was thus integrated into the heart of the worldview of the Yahad. It was not a marginal interest which could be easily concealed (as can be said for example about astrology in *b. šabbat* 156a-b), but rather a central tool in the anthropology of the Yahad. This notion is further exemplified below in a more detailed study of some aspects of 4Q186.

The Yahad's interest with scientific themes constitutes a significant advance from the apocalyptic interest in nature, which preceded the Yahad. In early Apocalypticism and in the contemporary wisdom traditions the study of nature played a significant role

[32] Baumgarten, "Qumranic and Astrological Terminology," 325; Philip S. Alexander, "Physiognomy, Initiation and Rank in the Qumran Community," in *Geschichte-Tradition-Reflexion: Festschrift für Martin Hengel zum 70. Geburtstag* (ed. P. Schaefer; Tübingen: Mohr Siebeck, 1996), 385-394. Popović is reluctant to accept this use of astrology in the Yahad, as he considers the text 4Q186 to have been written outside the sectarian context; however, he concedes that at some stage 4Q186 was indeed used in a sectarian context: see Popović, *Reading the Human Body*, 237-239.

[33] Translation follows Joseph M. Baumgarten & Daniel R. Schwartz, "Damascus Document (CD)", in *Damascus Document, War Scroll, and Related Documents* (ed. J.H. Charlesworth; DSSHAGT 2; Tübingen / Louisville, Mohr Siebeck / John Knox, 1995), 55.

alongside the revelations of past and future history.[34] The human need to catalogue the marvels of nature and reflect on them was *mutatis mutandis* conceived as a fertile mirror of the rules of nature for the eschatological age. Despite the famous prohibitions in Ben-Sirah (3:19-21) against revealed wisdom, it now seems probable that the difference between Ben-Sirah and the apocalyptic authors with regard to scientific themes is not as great as previously imagined, with the difference probably being that Ben-Sirah assigned a smaller role to revelation in comparison with his apocalyptic compatriots.[35] Other wisdom authors expand on cosmological themes, as can be seen in also some passages from the *Wisdom of Solomon*.[36] The acceptance of scientific themes in the Yahad thus rests on the solid ground of previous ideological trends. However, the Yahad does not directly continue any previous tradition, neither in terms of its general ideology nor in its relation to scientific themes.[37] Yahad literature uses the scientific and cosmological knowledge inherited from previous

[34] John J. Collins, "Cosmos and Salvation: Jewish Wisdom and Apocalyptic in the Hellenistic Age," *HR* 17 (1977): 121-142; Hartman, *Asking for a Meaning*; Stone, "The Parabolic Use of Natural Order" ; idem, "Lists of Revealed Things in the Apocalyptic Literature," in *Magnalia Dei - the Mighty Acts of God; Essays on the Bible and Archaeology in Memory of G. Ernest Wright*, (eds. F.M. Cross, W.E. Lemke and P.D. Miller; Garden City, NY: Doubleday, 1976), 414-452. More recently Klaus Koch, "The Astral Laws as the Basis of Time, Universal History, and the Eschatological Turn in the Astronomical Book and the Animal Apocalypse of 1 Enoch," in *The Early Enoch Literature* (eds. G. Boccaccini and J.J. Collins; JSJSup 121; Leiden: Brill, 2007), 119-137.

[35] Despite the assertions in 3:19-21, Ben-Sira does elsewhere treat revelation as his source of inspiration and knowledge. In addition, it seems that Ben Sira and Enoch are not too far apart, and probably belonged to the same circles in terms of their encyclopedic knowledge and interest in cosmology; see Annette Y. Reed, *Fallen Angels and the History of Judaism and Christianity: The Reception of Enochic Literature* (Cambridge: Cambridge University Press, 2005), 43-44; Benjamin G. Wright, "1 Enoch and Ben-Sirah: Wisdom and Apocalypticism in Relationship," in *The Early Enoch Literature*, (eds. G. Boccaccini and J.J. Collins; JSJSup 121; Leiden: Brill 2007), 159-176. *Contra* Randal A. Argall, *1 Enoch and Sirach: A Comparative and Conceptual Analysis of the Themes of Revelation, Creation, and Judgment* (Atlanta, Scholars Press, 1995), 74-76.

[36] John J. Collins, "The Reinterpretation of Apocalyptic Traditions in the Wisdom of Solomon," in *Jewish Cult and Hellenistic Culture. Essays on the Jewish Encounter with Hellenism and Roman Rule* (JSJSup 100; Leiden: Brill, 2005), 153-157, 158-180 (with earlier proponents of the same opinion quoted there) considers the *Wisdom of Solomon* to be the product of apocalyptic influence. Compare the different opinion by Reed, in the present volume.

(Aramaic) Wisdom and Apocalyptic traditions, while modifying them to fit their new context. The integration of astronomy and astrology into the sectarian texture was so profound, that one can speak of actual scientific creativity, i.e. of new modes of scientific learning created for sectarian needs.

3. Ideals of Science in the Apocalyptic Tradition

When contemplating the formation of a scientific discipline at a given society, it is necessary to not only evaluate the phenomenology of science in that discipline—i.e. the methods and tools, the experiments and techniques employed—but also the epistemology of that discipline: what its assumptions are about the world and the way it corresponds to the scientific product; what is the justification for scientific enquiry, what is the source of authority behind the study of nature. To use the terms of Amos Funkenstein, we seek not only the ideas of science but also its ideals:[38]

> *Ideals* of science differ in many ways from *ideas* of science. They indicate how a scientific community imagines science as it ought to be if ever completed; they express the ultimate criteria of rationality of their time. ... (my italics, Jonathan B.)
>
> ... in a certain sense, all science, every scientific argument or procedure, has an ideal—and, if you wish, fictional—aspect to it. It is the ultimate justification why the historian of science ought to distinguish between ideals and actual arguments, and then detect the former even in the latter.

[37] One should note, therefore, that the interest in scientific themes is an apocalyptic feature which continued—albeit with some modification—in Yahad ideology. This feature remains unnoted in the survey by John J. Collins, *Apocalypticism in the Dead Sea Scrolls* (London and New York: Routledge, 1997), even in the section titled "The heavenly World" (pp. 130-149). For the transformation of scientific theories as modifications rather than revolutions see Funkenstein, *Theology and the Scientific Imagination*, 12-18.

[38] Funkenstein, *Theology and the Scientific Imagination*, quotations from pp. 18, 21.

In modern manifestations of science the ideals of a given discipline would be such concepts as monocausality, or maybe more 'subjective' criteria such as harmony or symmetry or aesthetics, concepts that are often emphasized in recent histories of science. However, when discussing an ancient scientific discipline which left little explicit reflection on method (not so, by the way, in the study of Torah which was admirably reflective in the Yahad), the ideals of science should be formulated in a more simple way. I suggest the following three criteria which together constitute a sound epistemological basis for an ancient scientific discipline. The list is by no means definitive, as one may come forward with other constitutive elements, but I feel that my three criteria give a good account of the epistemic authority of science in the Dead Sea Scrolls.

The following principles are characteristic of the questions that a creative scientific discipline would ask itself:

> *a.* What are the origins of human knowledge about the world? The answer to this question would come in the form of a founding myth or an etiology, and quite often would invoke the name of a founding figure, a primary inventor or the like.[39]

> *b.* What is the justification for absorbing and accommodating earlier knowledge? This question will arise whenever earlier knowledge is embraced, for example with the circulation of Greek science and scholarly tradition among Islamic scholars.[40]

[39] A. Kleingünther, *PROTOS EURETES: Untersuchungen zur Geschichte einer Fragestellung* (Leipzig: Dietrich, 1933).

[40] See e.g. Dimitri Gutas, *Greek Thought, Arabic Culture: The Graeco-Arabic Translation Movement in Baghdad and Early 'Abbāsid Society (2nd-4th/8th-10th Centuries)*, (London: Routledge, 1998); Kevin T. van Bladel, *The Arabic Hermes: from Pagan Sage to Prophet of Science* (Oxford and New-York: Oxford University Press, 2009).

c. How does knowledge about the world function within the religious beliefs and practices of society? One may expect to see local narratives about the justification and use of science,[41] which would merge it within the fabric of society.

These characteristics will be investigated first as regards the apocalyptic milieu of the Enoch and Levi literature in the 2-3 centuries BCE, and subsequently with regard to the environment of the Yahad. Similarities and differences between the two milieus are pursued in order to highlight the unique scientific venture in the Yahad.

The early apocalyptic movement was a wide-scale project aiming to unveil the mysteries of the world by means of contemplating the revelations of primeval visionaries.[42] While strongly emphasizing the revelation of the future occurrences or the end of days, this movement also found great interest in revealing the secrets of ancient history, such as the tale of the Fallen Angels.[43] It also gave pride of place to

[41] See e.g. Simo Parpola, "Mesopotamian Astrology and Astronomy as Domains of the Mesopotamian 'Wisdom'," in *Die Rolle der Astronomie in den Kulturen Mesopotamiens* (ed. H.D. Galter; Grazer morgenländlische Studien 3; Graz: GrazKult, 1993); Maren Niehoff, "Inscribing Jewish Culture into Nature," *Philo on Jewish Identity and Culture* (TSAJ 86; Tübingen: Mohr Siebeck, 2001), 241-266.

[42] For the early Jewish apocalyptic tradition see John J. Collins, *The Apocalyptic Imagination: An Introduction to Jewish Apocalyptic Literature* (Grand Rapids, Mi.: Eerdmans, 1998); James C. VanderKam, *From Revelation to Canon: Studies in the Hebrew Bible and Second Temple Literature* (Leiden: Brill 2002); Andreas Bedenbender, *Der Gott der Welt tritt auf dem Sinai. Enstehung, Entwicklung und Funktionsweise der frühjüdischen Apokalyptik* (ANTZ 8; Berlin: Institut Kirche und Judentum, 2000). For the early Enoch literature see James C. VanderKam, *Enoch and the Growth of an Apocalyptic Tradition* (CBQMS 16; Washington D.C.: Catholic Biblical Association of America, 1983); Gabriele Boccaccini and John J. Collins (eds.), *The Early Enoch Literature* (JSJSup 121; Leiden: Brill, 2007). For the origins of Levi literature see Michael E. Stone, "Enoch, Aramaic Levi and Sectarian Origins," *JSJ* 19 (1988): 159-170; Robert Kugler, *From Patriarch to Priest: The Levi-Priestly Tradition from Aramaic Levi to Testament of Levi* (Atlanta: Scholars Press, 1996); Henryk Drawnel, *An Aramaic Wisdom Text from Qumran: a New Interpretation of the Levi Document* (JSJSup 86; Leiden: Brill, 2004).

[43] On the motivation of the urge to study primordial history see Ed Greenstein, "The Retelling of the Flood Story in the Gilgamesh Epic," in *Hesed Ve-Emet: Studies in Honor of Ernst S. Frerichs* (eds. J. Magness and S. Gitin; BJS 320; Atlanta: Scholars Press), 197-204; and quite differently John van Seters, *Prologue to History: The Yahwist as Historian in Genesis* (Westminster:

studying the secrets of the created world. Apocalypticism believes that the Divine Order of the world becomes manifest in both the temporal and spatial aspects of the cosmos, indeed that they can all be fit into one harmonious entity. Therefore a true disciple of this movement is compelled to study the marvels of the Universe on the manner of the great patriarch Enoch (cf. 1 En 82:1-4a, 93:11-14). The parade example for this kind of wisdom is the Book of Astronomy, which was considered to be part of the Enochic heritage already at an early stage of the tradition.[44]

a. The Founding Myth: Origins of Knowledge

The *Book of Watchers* is an early Enochic composition which now covers chapters 1-36 of 1 Enoch. It tells of the grave sins of the Fallen Angels, who have taken the daughters of men, as depicted in Genesis 6:1-4 and expanded in the Enoch tradition. This myth in many ways becomes the constitutive myth of the apocalyptic tradition.[45] But the Watchers, as they are often called, were not only guilty of perpetrating sex and violence, but rather also of introducing to humanity the secrets of arts, crafts, and knowledge. In a Promethean mode of thought, the writer (or better writers, as there are most probably several intertwined traditions) reports:[46]

John Knox Press, 1992).

[44] For the incorporation of the Book of Astronomy in the Enoch tradition see George W. E. Nickelsburg, *1 Enoch 1: A Commentary on the Book of 1 Enoch* (Hermeneia; Minneapolis: Fortress, 2001), 21-26; Koch, "The Astral Laws". Much of the reasoning should change according to the finds in R.D. Chesnutt, "Oxyrhynchus Papyrus 2069 and the Compositional History of 1 Enoch," *JBL* 129 (2010): 485-505. Apparently we now have a Greek copy of the Book of Astronomy which is embedded with other Enochic compositions.

[45] On the traditions of the Fallen Angels see: Devorah Dimant, *"The Fallen Angels" in the Dead Sea Scrolls and in the Apocryphal and Pseuepigraphic Books Related to them* (Dissertation, Hebrew University of Jerusalem, 1974) in Hebrew; George W.E. Nickelsburg, "Apocalyptic and Myth in 1 Enoch 6-11," *JBL* (1977): 383-405; C. Auffarth and L.T. Stuckenbruck, *The Fall of the Angels* (TBN 6; Leiden: Brill, 2004); Archie T. Wright, *The Origin of Evil Spirits: The Reception of Genesis 6.1-4 in Early Jewish Literature* (WUNT n.s. 198; Tübingen: Mohr Siebeck, 2005); Reed, *Fallen Angels*; Veronika Bachmann, *Die Welt im Ausnahmezustand: eine Untersuchung zu Aussagegehalt und Theologie des Wächterbuches (1 Hen 1-36)*, (BZAW 409; Berlin: de Gruyter, 2009).

[46] Translation follows Nickelsburg, *1 Enoch 1*.

> 8:1. Asael taught men to make swords of iron and weapons and shields and breastplates and every instrument of war. He showed them metals of the earth and how they should work gold to fashion it suitably, and concerning silver, to fashion it for bracelets and ornaments for women. And he showed them concerning antimony and eye paint and all manner of precious stones and dyes... 3. Shemihazah taught spells and the cutting of roots. Hermani taught sorcery for the loosing of spells and magic and skill. Baraqel taught the signs of the lightning flashes. Kokabel taught the signs of the stars. Ziqel taught the signs of the shooting stars. Artoqeph taught the signs of the earth. Shamshiel taught the signs of the sun. Sahriel taught the signs of the moon.

The transmission of knowledge to humanity—technology, divination, crafts, astronomy, astrology—is thus a grave sin. In contrast, other Enochic compositions trace a more benign source for human knowledge, as in the introduction to the Book of Astronomy.[47] Here it is Enoch who taught knowledge:

> 72:1 The book about the motion of the heavenly luminaries, all as they are in their kinds, their jurisdiction, their time, their name, their origins, and their months which Uriel, the holy angel who was with me (and) who is their leader, showed me.

[47] The contradiction between the positions of BW and AB is apparent in the present literary content of *1 Enoch*, less so in earlier manifestations of the anthology of Enochic booklets, where it was not clear that AB and BW belong to the same collection. However, the contradiction was evidently conceived by the author of Jubilees, who produced the image of Enoch in 4:17-21 based on his knowledge of previous Enochic traditions, while also being sensitive to questions on the origin of knowledge. See Michael A. Knibb, "Which Parts of 1 Enoch Were Known to Jubilees? A Note on the Interpretation of Jubilees 4.16-25," in *Reading from Right to Left: Essays on the Hebrew Bible in Honour of David J.A. Clines* (eds. J.C. Exum and H.G.M. Williamson; JSOTSup 373; Sheffield: Academic Press, 2003), 254-262; John S. Bergsma, "The Relationship between Jubilees and the Early Enochic Books (*Astronomical Book* and *Book of Watchers*)", in *Enoch and the Mosaic Torah: The Evidence of Jubilees* (eds. G. Boccaccini and G. Ibba; Grand Rapids, Mi.: Eerdmans, 2009), 36-51.

> 81:2 I looked at all the heavenly tablets, read everything that was written, and understood everything. I read the book of all the actions of people and of all humans who will be on the earth… 5. Those seven holy ones brought me and set me on the earth in front of the gate to my house. They said to me: "Tell everything to your son Methuselah and show all your children that no human is righteous before the Lord… 6. We will leave you with your son for one year until you again live your (last?) command, to teach your children, write for them, and testify to all your children. In the second year they will take you from them.

A number of traditions about the sins of the watchers are collected in 1 Enoch 7-8 in a somewhat haphazard way.[48] While a basic version of the myth disclosed the acts of violence performed by the angels, other traditions emphasize that the transgression consisted—either in addition or alternatively—in the transmission of knowledge. A first list concentrates on technological skills related with sex and violence: not only did the Watchers indulge in these two vices, they also revealed to mankind some material qualifications used to that end:[49] the mining and production of metals for the production of arms and jewelry, as well as the mining and production of cosmetics. Yet another literary layer introduces a list of 'applied' sciences revealed by the sinning angels. Most of this material reflects the combination of divinatory and magical acts which are characteristic of Mesopotamian scholarship and *āšipūtu*: cutting roots, absolving spell-charms, as well

[48] Dimant, *Fallen Angels*, 52-65; Nickelsburg, "Apocalyptic and Myth"; Carol A. Newsom, "The Development of Enoch 6-19: Cosmology and Judgment," CBQ 42 (1980): 310-329; Reed, *Fallen Angels*, 27-44; Helge Kvanvig, *Primeval History: Babylonian, Biblical, and Enochic. An Intertextual Reading* (JSJSup 149; Leiden: Brill, 2011), 453-469; Henryk Drawnel, *The Aramaic Astronomical Book from Qumran: Text, Translation, and Commentary* (Oxford and New-York: Oxford University Press), 53-70. Dimant, Nickelsburg and Newsom describe the *Book of Watchers* as an accretion of two or three different traditions, as accepted also by Michael Segal, *The Book of Jubilees: Rewritten Bible, Redaction, Ideology and Theology* (JSJSup 117; Leiden: Brill, 2007). More recently, Nickelsburg, *1 Enoch 1*, 171-172 speaks of "a series of expansions, elaborations or accretions".

[49] Nickelsburg, *1 Enoch 1*, 171.

as the arts of deducing portents from the heavenly luminaries, as taught in the cuneiform series *Enūma Anu Enlil*.[50] This use of the Watchers' teaching is seen also in the terminology נ]חשי כוכבין and נחשי שמ]ש, 'au]guries of stars' and 'auguries of the su[n' with the word נחש indicating a clear divinatory meaning, which is not evident in the various Greek and the Ethiopic translation of this passage.[51] In BW knowledge about the luminaries is depicted as a part of illicit divination, while AB presents a perfectly legitimate version of it.

While the wisdom of the watchers is depicted as negative and sinful, the transmission of Enoch's revelation seems positive and benign. The difference is at least partly due to the magical application of the watchers' knowledge, which is nothing but witchcraft (φαρμακεία) in the account of BW (1 En 8:3, Syncellus). The message is thus an anti-magical one, and can be evidently seen in the AB, where an abstention from astrology and astrological themes is clearly discerned.[52] Enoch's knowledge in AB is more theoretical and lacks an applied aspect.

Alexander conceived of the difference between the wisdom of the Watchers and that of Enoch through a prism of national identity: while the Watchers represent alien wisdom, Enoch represents the Jewish tradition.[53] The alien source in this hypothesis could be Greek, as suggested for example by David Suter, or, more probably, Mesopotamian. Recently, Henryk Drawnel and others claimed for a full correspondence between the teaching of the Watchers and the practices of the traditional Mesopotamian *āšipu* and *ṭupšar Enūma Anu Enlil*.[54] The famous fragment from Pseudo-Eupolemos about Enoch and Abraham as the inventors of astrology vis-à-vis the Greek sages proves

[50] Drawnel, *Aramaic Astronomical Book*, 53-70.

[51] Michaël Langlois, *Le premier manuscrit du livre d'Hénoch: Étude épigraphique et philologique des fragments araméens de 4Q201 à Qumrân* (Paris: Cerf, 2008), 259.

[52] Matthias Albani, *Astronomie und Schöpfungsglaube: Untersuchungen zum astronomischen Henochbuch* (WMANT 68; Neukirchen-Vluyn: Neukirchener Verlag, 1994), 335-344; Reimund Leicht, *Astrologumena Judaica: Untersuchungen zur Geschichte der astrologischen Literatur der Juden* (TSMJ 21; Tübingen: Mohr Siebeck, 2006), 28-35; *contra* Christfried Böttrich, "Astrologie in der Henochtradition," *ZAW* 109 (1997): 222-245.

[53] Alexander, "Enoch and the Beginnings of Jewish Interest in Natural Science," in the present volume. See the criticism by VanderKam, in the present volume.

that the figure of Enoch was indeed used for this purpose, of instituting a Jewish culture hero. The question is, however, whether the Enochic authors were aware that their scientific traditions are originally Mesopotamian, and whether indeed they meant to domesticate them by means of the Enochic stories. More specifically, the question is whether, as Drawnel claims, the Watchers are designed as a parody on Mesopotamian *ummanū*. The problem is of course that this opinion is never noted in the BW, in contrast for example to the explicit parody in the Book of Daniel. In addition, the figures of the watchers as primordial sages have much deeper roots in the mythological imagination than a mere parody on human protagonists. The scene of intellectual dispute between Israel and the gentiles is thus better anchored in Daniel than in BW, where other explanations yield better results.[55] Note that, while Daniel is clearly a Jewish hero, Enoch is not exactly a Jewish protagonist par excellence. His figure—both in the BW and in the *Book of Giants* (where there are some hints as to his whereabouts)—is more that of a liminal sage than a pious Israelite.[56]

If one seeks to legitimate the origins of science and technology among mankind, the Watchers are a rather bad choice to play the culture hero. However, we may gain a better glimpse of them by acknowledging the mythical character of the Watchers' story, a view which entails some complexity which could not have been expected in a more theologically-oriented text. It is an essential characteristic of ancient myths that mediating figures which stand between the human

[54] For the Greek hypothesis see David Suter, "Fallen Angel, Fallen Priest: The Problem of Family Purity in 1 Enoch 6-16," *HUCA* 50 (1979): 115-135. For the Mesopotamian option see Drawnel, *Aramaic Astronomical Book,* 54-73; cp. Kvanvig, *Primeval History,* 453-469; Amar Annus, "On the Origin of Watchers: A Comparative Study of the Antediluvian Wisdom in Mesopotamian and Jewish Traditions," *JSP* 19 (2010): 290-291. Although the work by Annus presents a bounty of sources and connections, their application to the Jewish and Aramaic traditions should be taken with a grain of salt.

[55] See Alan Lenzi, "Secrecy, Textual Legitimation, and Intercultural Polemics in the Book of Daniel," *CBQ* 71 (2008): 330-348. As well as the article by Sanders in the present volume.

[56] Newsom, *The Self as Symbolic Space*, 49; Jonathan Ben-Dov, "Hebrew and Aramaic Writing in the Pseudepigrapha and the Qumran Scrolls: The Ancient Near Eastern Background and the Quest for a Written Authority," *Tarbiz* 78 (2009): 27-60, here 38-49 (in Hebrew).

and divine world carry both benign and malevolent aspects. Thus, for example, the use of such protective creatures as *kārib(t)u*, *lamassu* in Assyrian and Babylonian palaces, and of *apkallu* in glyptic art.[57] Their agency is neither good or evil: they are powerful beings whose force lies beyond good and evil. The acts of these mediators may be apprehended by human beings as good or as evil in a particular circumstance, but it would be wrong to associate goodness or evil with them. The fusion of man, beast, and divine within a single creature breaks the normal cultural categories, creating an abnormality and necessarily raising disorder, awe and violence. Ambiguity in this case is a horrifying trait. This is true also in the Greek mythology, as in the case of Prometheus, the Titan, whose figure exemplifies the mediating figure *par excellence*. Being a Titan, he is *a priori* conceived as an enemy of the structured world, which later flourishes under the dominion of Zeus. Yet Prometheus conveyed to humanity the ability to control fire. Fire itself is an ambiguous symbol, being an emblem of technology and the domestication by means of industry, while on the other hand it is quite often the foremost example of uncontrollable disaster. In addition, Prometheus is ultimately responsible not only for the human mastery of technology, but also for the calamities brought by Pandora.

Mediation is thus a tricky business, with the figure of the Watchers being a prime example of this double-edged sword. While bestowing to humanity the great benefits of technology, they act with violence and terror and nearly cause the extermination of mankind. Recent publications redefine the image of the watchers on the backdrop of Mesopotamian *apkallu* traditions, where the ambiguity of the protagonists is discerned.[58] Depicted as a man-fish emerging from the primordial waters, or as a man dressed in fish skin, the *apkallu* is the perfect *Zwischenwesen*. Oannes, the first *apkallu*, is called by

[57] See for example Anthony Green, "Beneficent Spirits and Malevolent Demons: The Iconography of Good and Evil in Ancient Assyria and Babylonia," in *Popular Religion, Visible Religion 3* (ed. H.G. Kippenberg; Leiden: Brill, 1984), 80-105; Franz A.M. Wiggermann, *Mesopotamian Protective Spirits: the Ritual Texts* (CM 1; Groningen: STYX, 1992); Karen Sonik, *Daimon-Haunted Universe: Conceptions of the Supernatural in Mesopotamia* (Dissertation, University of Pennsylvania, 2010), 47-74.

[58] Annus, "On the Origins of Watchers"; Kvanvig, *Primeval History*, 107-158.

Berossus 'a senseless beast' (ζῷον ἄφρενον).[59] While previous scholarship stressed the positive parts of the tradition, more could be done with traditions such as those preserved in the series *bīt mēseri*, reporting the sins of the *apkallu*: "... who angered the god Adad in heaven.... Who hung his seal on a 'goat-fish' and thereby angered the god Enki\Ea...".[60] Within the Cuneiform tradition there are signs that the transmission of knowledge by the primordial sages was not always considered a benign act.[61] As Fritz Graf points out, technological inventions were also considered as 'ambivalent power' in the Greco-Roman tradition.[62] Part of the malign nature of the watchers is undoubtedly due to this mythological aspect. Enoch himself is a liminal figure just like the watchers, and his frequent associations with

[59] As Dr. Romina Vergari (Perugia and Haifa) kindly informs me, the word ἄφρενον is derived from α-φρην, i.e "without heart/mind/sense" (cp. Xenophon, *Mem* 1,4,4 εἴδωλα ἄφρονά 'senseless statue'). I see little justification for the translation in P. Schnabel, *Berossos und die babylonisch-hellenistische Literatur* (Leipzig: Teubner, 1923), 253 "ein furchtbares Untier"; as well as Gerald Verbrugghe and John M. Wickersham, *Berossos and Manetho, Introduced and Translated: Native Traditions in Ancient Mesopotamia and Egypt* (Ann Arbor: University of Michigan Press, 1996), 44 "a frightening monster".

[60] Quoted from Kvanvig, *Primeval History*, 108-109, cf. Annus, "On the Origins of Watchers," 297-298. The main problem arising from this passage is the apparent ambiguity, even contradiction, between the image of the *apkallū* as mythological figures in ritual and their appearance as living human beings in scholarly texts; on this problem see the debate between Sanders and Lenzi *apud* Alan Lenzi, *Secrecy and the Gods: Secret Knowledge in Ancient Mesopotamia and Biblical Israel* (SAAS 19; Helsinki: Helsinki University Press, 2008), 108-113.

The recurrent invocation of *apkallū* and use of their images in magical and apotropaic contexts suggests, according to Annus, that these primordial figures have a certain demonological dimension. However, as Seth Sanders kindly informs me, this claim is going too far, since obviously the *apkallū* are called to scare away the demons, not to help them (*Utukkū lemnūtu* 7:109; Mark J. Geller, *Evil Demons: Canonical Utukkū Lemnūtu Incantations* [SAACT 5; Helsinki: Helsinki University Press, 2007], 140, 223.)

[61] Andrew R. George, *Babylonian Literary Texts in the Schøyen Collection* (Bethesda, MD: CDL Press, 2009), 110; cf. the review of this work by Alan Lenzi, *RBL* 01/2011.

[62] Fritz Graf, "Mythical Production: Aspects of Myth and Technology in Antiquity," in *From Myth to Reason? Studies in the Development of Greek Thought* (ed. R. Buxton; Oxford: Oxford University Press, 1999), 317-328. The ambiguity of astral divination and other arcane knowledge is most apparent in the Book of *Zohar*; see Yehuda Liebes, *The Cult of the Dawn. The Attitude of the Zohar towards Idolatry* (Jerusalem: Carmel, 2011, in Hebrew).

them in those primordial days placed him too in that numinous realm standing 'betwixt and between' the cultivated world and the world of the demons. In this respect, both associations of the origin of knowledge—to Enoch and to the watchers—convey the message that natural science has revered origins, and is thus to be both respected and feared at the same time.

The distinction between the Watchers' wisdom and that of Enoch also involves the extent of distribution of cosmological wisdom. The ambiguity within Enochic literature with regard to the value of science —whether sinful or benign—represents the question how much it is legitimate to distribute esoteric teaching in public. Mesopotamian science, which served in some way as the ancestor of Enochic science, practiced strict limitations on the distribution of knowledge outside the circles of the initiated. These limitations took the literary form of short formulary prohibitions incorporated in the colophons of scientific texts.[63] The variant evaluation of astral lore in Enochic literature may thus be due to the degree of the prohibition on communicating secret knowledge: while Enoch and his progeny are considered legitimate transmitters of esoteric wisdom, the transmission by the Watchers was illegitimate as it involved both dubious teachers and incompetent students.[64]

The association of the Watchers with the sin of transmitting knowledge, so prominent in the *Book of Watchers*, loses its popularity and gradually disappears from subsequent Enoch traditions. Thus, in Jubilees for example, the sin of the watchers is limited to violence and sex, and the discovery of sciences and writing is assigned to Enoch alone (4:17-21). A similar situation pertains in later literary sources such as the early Church fathers, some notable exceptions being the *Book of Parables* (Chapter 69), Pseudo-Philo 34:3, and the Christian Orthodox tradition, which retain the old themes of the angelic

[63] Mladen Popović, "Physiognomic Knowledge in Qumran and Babylonia: Form, Interdisciplinarity, and Secrecy," *DSD* 13 (2006): 150-176. Lenzi, *Secrecy and the Gods.* For secrecy in earlier cuneiform tradition see Joan Goodnick-Westenholz, "Thoughts on Esoteric Knowledge and Secret Lore," in *Intellectual Life of the Ancient Near East: Papers Presented at the 43rd Rencontre Assyriologique Internationale, Prague, July 1-5, 1996* (Prague: Oriental Institute, 1998), 451-462.

[64] See Reed, in this volume.

transmission of illicit knowledge.[65] One way to explain the disappearance of forbidden knowledge from the Jewish Enochic tradition is that the mythological ambiguity of the mediating figures was not well-accepted by later apocalyptic literature, where a clearer black-and-white polarity prevailed: Enoch = good; Watchers = bad. In addition, the fact that the malevolent origin of science and technology gradually gives way to a benign view of these disciplines proves that the legitimacy of scientific speculation was growing. What had functioned in earlier times as an apologetic argument in favor of the scientific reflection no longer required apologetics, since by then the sciences were safely backed by the legitimate figure of Enoch, both in the Book of Astronomy (1 En 82:1-3) and in Jubilees 4:17-21.[66]

Finally, the metamorphosis of cosmological knowledge from forbidden revelation to pious discipline may be due to the accommodation of the—originally Mesopotamian—outlook on science into the prevalent attitude in Hellenistic Judea. In fact, at that time even the Babylonian environment itself was heavily Hellenistic, as can be discerned both from mythological accounts (Berossus) and divinatory-scientific texts.[67] Populations of the Hellenistic Levant happily joined forces with the Hellenistic quest after a *prōtos heuretēs*, promoting the Antiquity of their national traditions by positing a primordial culture-hero. This motif is often invoked by Phoenician-Hellenistic writers and appears also in the heavily-studied passages from the Jewish authors Pseudo-Euploemos, Josephus, the

[65] Dimant, *Fallen Angels*, 180-181; Graf, "Mythical Production," 321-322; Reed, *Fallen Angels*, 160-189.

[66] Curiously, Jubilees and other authors assign the transmission of primordial knowledge to the sons of Seth. See Albertus F.J. Klijn, *Seth in Jewish, Christian and Gnostic Literature* (NovTSup46; Leiden: Brill, 1977); Andrei A. Orlov, "Overshadowed by Enoch's Greatness: 'Two Tablets' Traditions from the *Book of Giants* to Paleia Historica," *JSJ* 32 (2001): 137-158.

[67] Examples from divination: JoAnn Scurlock and Farouq N.H. Al-Rawi, "A Weakness for Hellenism," in *If a Man Builds a Joyful House. Assyriological Studies in Honor of Erle Verdun Leichty* (ed. A.K. Guinan; CM 31; Leiden: Brill, 2006), 357-382; JoAnn Scurlock, "Sorcery in the Stars: STT 300, BRM 4.19-20 and the Mandaic Book of the Zodiac," *AfO* 51 (2006): 125-146. It was claimed that two late cuneiform texts dealing with shadow length, or, more likely, manuals for the constructions of sundials, are written using Greek methods: Francesca Rochberg-Halton, "Babylonian Seasonal Hours," *Centaurus* 32 (1989): 146-170.

Genesis Apocryphon from Qumran, etc.[68] What had originally echoed the ambiguous origins of wisdom in one mythological tradition gave way to a more unequivocal assertion of the value of science in a later, more Hellenistic-oriented, setting.[69]

The motivation to study scientific themes exemplifies the commitment of apocalyptic writers to the exemplary revelation of revered patriarchs. The one-time revelation given to Enoch, Levi, Isaac, or whoever else, was not expected to reoccur to any of the initiated. The wisdom of Enoch is *prisca theologia*: the one true teaching to which all future practitioners should attune. There are very few reflective statements in this literature, i.e., statements that prompt the reader to contemplate on it and study this wisdom further. In *1 En* 81:1-3, one of few such passages, Enoch is commanded to write his vision in books and transmit them to future generations. Other passages found in the Aramaic texts from Qumran treat apocalyptic wisdom in general rather than the scientific themes in particular. In these passages wisdom is an 'inheritance', its distribution limited to a closed circle with strict restrictions on strangers and foreigners (4Q542 Testament of Qahat ar 1 i 4-6).[70] This wisdom is not renewable but rather conservative; it is content in preserving the old teaching rather than updating it. The preservation of this wisdom is the right way to salvation and the ultimate implementation of wisdom in the eschatological age.

[68] Gerorge H. van Kooten, "Enoch, the 'Watchers', Seth's Descendants and Abraham as Astronomer. Jewish Applications of the Greek motif of the First Inventor (300 BCE - CE 100)", in *Recycling Biblical Figures: Papers Read at a NOSTER Colloquium in Amsterdam 12-13 May* 1997 (eds. J.W. van Henten and A. Brenner; Leiderdorp: Deo Publishing, 1999), 292-316; Annette Y. Reed, "Abraham as Chaldean Scientist and Father of the Jews: Josephus, Ant. 1.154-168, and the Greco-Roman Discourse about Astronomy/Astrology," *JSJ* 35 (2004): 119-158.

[69] For the changing perspectives of the Enoch tradition in the Hellenistic period see Annette Y. Reed, "The Origins of the *Book of the Watchers* as 'Apocalypse' and its Reception as 'Apocrypha,'" *Henoch* 30 (2008): 55-59.

[70] See Ben-Dov, "Scientific Writings," 397-398. This revealed knowledge may be transmitted either by means of a chain of ancient books or by oral transmission of a primordial revelation; see Eibert J.C. Tigchelaar, "Aramaic Texts from Qumran and the Authoritativeness of Hebrew Scriptures: Preliminary Observations," in *Authoritative Scriptures in Ancient Judaism* (ed. M. Popović; JSJSup 141; Leiden: Brill, 2010), 170-171.

b. Reception and Accommodation

For us modern scholars it seems clear that the Enochic science is in many ways an inheritance of earlier, mostly Mesopotamian traditions. This statement requires some supplementary qualifications, since evidently the Judean scientific discipline is not a mere branch of the Babylonian one but rather a new cultural formation, with its own aims, methods, and presuppositions. In addition it undoubtedly embraced also other, non-Mesopotamian, concepts and molded them together with the Mesopotamian infrastructure.[71] Nevertheless, the Mesopotamian origin of some basic concepts is hard to deny. The reliance is especially apparent in the lunar visibility texts of *Enūma Anu Enlil* tablet XIV, as converted into the Aramaic medium in 4Q208 and 4Q209 (or more probably the forerunner thereof).[72] The question remains, however, whether or not the adherents of the emerging Jewish scientific discipline were aware of the foreign origin of their teaching, and what they did to incorporate it in the authoritative framework of the Jewish literary creation.

The literary form of the early apocalyptic scientific lists is instructive in this regard, as it discloses the authoritative voice chosen by the authors. Both the *Astronomical Book* and the *Aramaic Levi Document* are the products of the third, or possibly the early second century BCE, and are framed by a narrative about patriarchal times, given mostly in the first person of Enoch or Levi. The 'I' of the speaker is the patriarch himself, who lends his authority to the scientific material included in his book.[73] The technical materials, which originally had taken the form of a technical roster or table without any particular author, now appear as the testament of a patriarch to his sons, sometimes even as part of a creative dialogue. Enoch and Levi are thus depicted as the proto-scientists, who lend credibility to the entire scientific enterprise. Somewhat similarly, a later apocalyptic text like

[71] See the contributions of Popović and Yoshiko Reed in this volume.

[72] Drawnel, *Aramaic Astronomical Book*; idem, "Moon Computation in the Aramaic Astronomical Book," *RQ* 23 (2007): 3-41; Ben-Dov, *Head of all Years*, 169-174, 189-192.

[73] This strategy is characteristic of the Aramaic texts, including the *Genesis Apocryphon*; see Ben-Dov, "Hebrew and Aramaic Writing in the Pseudepigrapha and the Qumran Scrolls"; Tigchelaar, "Aramaic Texts from Qumran."

Fourth Ezra depicts the interest of Ezra with cosmology, and a Jewish-Hellenistic text like the *Wisdom of Solomon* which was possibly influenced by apocalyptic thought, would rely on the firm authority of King Solomon, the wise of all men.

c. Religious function of science

Science in the early Enoch compositions is part of the Eschatological *Weltbild*. It is the received wisdom that should be learned when uncovering the heavenly mysteries, in order to unveil the Divine plan and improve one's ability to survive when the final day comes. More simply, knowing the correct order of the world is necessary in order to understand what happens when the world deviates from its norm, a common apocalyptic trope. A classical text for this purpose is 1 Enoch 80:2-8, a passage which seems to be a late part of the editorial framework of the book:

> In the days of the sinners the years will grow shorter. Their seed will be late on their land in their fields. Everything on earth will change and will not appear at their times... The moon will change its order and will not appear at its (normal) time... Many heads (i.e., leaders, cf. 82: 11-13, JBD) of the stars will stray from the command... The entire law of the stars will be closed to the sinners...

In this passage, the natural order is not distinct from the ethical or covenantal order. Carrying an ancient biblical notion to its extreme, this author proclaims the essential unity of *physis* and *nomos*. If one learns the astronomical basics of 1 Enoch 72-79, he will be sure to notice the deviations from Nature which materialize in the age of the sinners.[74]. For this purpose it would suffice to rehearse the ancient

[74] For the biblical unity of nature and law see Hans H. Schmid, *Gerechtigkeit als Weltordnung: Hintergrund und Geschichte der alttestamentlichen Gerechtigkeitsbegriffes* (Tübingen: Mohr Siebeck, 1968). VanderKam, in his article in the present volume, stresses the incompatibility of chapter 80 with the main bulk of AB with regard to the stability of the laws of nature: the possibility that Nature will be disrupted contradicts the regularity of Nature according to the rest of AB. However, I do not find this difference compelling. On the contrary, the schematic and highly idealized character of Enochic science necessarily requires explanations for the discrepancy between these schemes and the less regular nature. The gap between the regularity of nature at Creation and the corruptness of the present-day

teachings inherited from Enoch and Levi, without developing them further or practicing any scientific creativity.

4. The Scientific Environment within the Yahad

As demonstrated in some detail in section 2 (above), scientific themes such as astronomy and astrology were carried within the Yahad further than their employment in preceding sapiential and apocalyptic traditions. It is now time to reexamine these scientific notions in Yahad literature in light of the analysis presented so far.

a. Modes of Acquiring Knowledge

A fundamental point distinguishes the mode of revelation celebrated in the Yahad from that of previous apocalyptic trends: the Yahad allows for less strict learning procedures and encourages new learning. While this point is made explicit only with regard to the study of Torah, we have reason to believe that it applies to the study of the natural world as well. It is a potential catalyst for the Yahad's attitude to new ventures in scientific themes. This conception is demonstrated in some passages of columns 8-9 of *Serekh ha-Yahad*.

1QS IX 12-14[75]

> אלה החוקים למשכיל להתהלכ בם עם כול חי לתכון עת
> ועת ולמשקל איש ואיש לעשות את רצון אל ככול הנגלה
> לעת בעת ולמוד את כול השכל הנמצא לפי העתים ואת
> חוק העת

> These are the statutes, by which the Maskil shall walk with every living being, according to the measure of each time and the weight of each man, to do God's will according to everything which has been revealed from

world is built into the apocalyptic worldview and explains ideas such as "the New Creation," so prevalent in apocalyptic writings (1 Enoch 72:5, Jub 1:29 etc.). Such a concept is reflected to a certain extent in Mesopotamian texts like the literary framework to the lunar section of *Enūma Anu Enlil*; see Ben-Dov, *Head of all Years*, 205-207, and previous literature cited there. There are other grounds why chapter 80 is not part of the original AB, but the above mentioned theological argument does not rule the case entirely.

[75] Translation follows Devorah Dimant, "Time, Torah and Prophecy at Qumran," in *Religiöse Philosophie und philosophische Religion der frühen Kaiserzeit* (eds. R. Hirsch-Luipold, H. Goergemanns and M. von Albrecht; STAC 51; Tübingen: Mohr Siebeck, 2009), 155-156.

> time to time, and to measure the understanding which has been found according to the times, and the law of the time.

As Dimant indicates, "the most striking feature of this remarkable statement is its temporal perspective".[76] The passage conveys the clear notion that the revelation of God's mysteries is progressive and constantly susceptible to modification, both with regard to "the measure of each time" and "the weight of each man". The study of Torah is thus a dynamic process, and new revelations are available to the members of the Yahad at any time—given that they remain loyal to the Yahad's way of life. A similar concept of progressive revelation is invoked earlier on in the Serekh, aligning the modes of interpretation of the words of the prophets with those of the Torah (1QS VIII 15-16). This notion contrasts with the notion of a one-time antiquarian revelation, as attested in the apocalyptic writings.[77] Not only the *Maskil* but also each member of the Yahad is eligible to receive such revelation, as in 1QS VIII 11-12:[78]

> וכול דבר ה̇נסתר מישראל ונמצאו לאיש הדורש אל
> יסתרהו מאלה מיראת רוח נסוגה

> No doctrine concealed from Israel but discovered by an interpreting man is to be hidden from these men out of fear that they might backslide.

This short statement opens a window to the learning process in the Yahad, by which it is made clear that not only the Teacher, but also lay members of the Yahad expected to receive a revelation

[76] Dimant, "Time, Torah and Prophecy," 156. While this aspect was already noted by Lawrence H. Schiffman, *The Halakhah at Qumran* (Leiden: Brill, 1975), 33-36 and earlier studies, Dimant made a great step forward by shedding light on this concept as part of a comprehensive conception of history within the Yahad. The passage was discussed recently also by Jassen, who suggests important distinctions between the notion of progressive revelation in the study of Torah (1QS IX) and the study of Prophecy (1QS VIII). Jassen presents a fuller discussion of the variants to this section in the parallel copy 4QS[e] in Alex P. Jassen, *Mediating the Divine: Prophecy and Revelation in the Dead Sea Scrolls and Second Temple Judaism* (STDJ 68; Leiden: Brill, 2007), 338-342.

[77] Dimant, "Time, Torah and Prophecy," 152.

[78] Translation follows DSSEL with my corrections.

enlightening a segment of the teaching.[79] Thus, alongside the progressive revelation, the Yahad's learning procedures were relatively open and only meagerly hierarchic.[80] This setting is far more institutionally better for study than the vague transmission modes of the Enochic literature. It also encourages a freer indulgence with older texts and ideas, rather than being limited to perpetuating them. This kind of atmosphere may shed light on the background for the way scientific concepts were treated within the Yahad.[81]

While the *Serekh* passage quoted here refers specifically to the interpretation of the Torah, there are reasons to believe that scientific

[79] See also Albert I. Baumgarten, "Information Processing in Ancient Jewish Groups," in *Sectarianism in Early Judaism: Sociological Advances* (ed. D.J. Chalcraft; London: Equinox, 2007), 246-255. The DSSEL translation understands the term האיש הדורש as "The Interpreter," referring to one specific interpreter in the past, as in the figure of דורש התורה in CD. However, the term איש דורש is never applied to that past figure, but rather refers to a layman, as is suggested in my translation. This understanding is also shared by Eyal Regev, "Between Two Sects: Differentiating the Yahad and the Damascus Covenant," in *The Dead Sea Scrolls: Texts and Context* (ed. C. Hempel; STDJ 90; Leiden: Brill, 2010), 437 n.21; cp. idem, "The Yahad and the Damascus Covenant: Structure, Organization and Relationship," *RQ* 21 (2003): 233-262.

[80] The progressive revelation and the learning procedures have been causes for disagreement among scholars lately. Regev in a series of articles suggested that while both traits are characteristic of the Yahad, they are not characteristic of the community described in the *Damascus Document*. This latter community, says Regev, denied the possibility of progressive revelation and imposed a strict hierarchic order on its learning members. However, most scholars view the progressive revelation of the D community as equivalent to the Yahad based on such statements as CD XII, 20-22 (Dimant, "Time, Torah and Prophecy," 159), as well as XV, 10 אל הנמצא לעשות בכל קץ קרבו 'that which is found to be done at any period of time in which he approaches (the community)' (Menahem Kister, private communication). The question of the availability of revelation to laymen is not directly addressed in D, and thus remains to be decided on the basis of Regev's other evidence. Generally speaking, Regev's distinction between the D community and the S community gained only partial support: according to Collins, *Beyond the Qumran community*, 48-65 the differences are not compelling enough to consider them two separate communities. For the present discussion I limit myself to the Yahad without a clear conviction on the D community.

[81] Martin Hengel, "Qumran and Hellenism," 51-55 sees the "intellectual" atmosphere of the Yahad (which he detects in other quotes and examples) as the product of a Hellenistic mode of thought. While Hellenism is indeed a possible source, other factors may also be suggested too from within the Jewish tradition.

knowledge, especially diagnostics by means of astrology, was considered according to this epistemological model too. In the same passage from 1QS IX 12-14 quoted above, the teaching of the Maskil is qualified "according... to the weight of each man". This component is clarified in the immediately following lines 14-15:

להבדיל ולשקול בני הצדוק (4QS^e הצדק) לפי רוחם...
ואיש כרוחו כן לעשות משפטו

> ...to separate and weigh the Sons of Righteousness according to their spirit... and according to a man's spirit (is) justice to be done (to him)[82]

The diagnosis (weighing) of community members according to their spirit is thus the Maskil's duty alongside the progressive interpretation of the Torah. This vocation is depicted in other descriptions of the Maskil, as in 1QS III 13-15 (quoted above). As noted above, this passage presents several diagnostic terms which were part of the science of the day: תולדות ('character'),[83] רוח ('spirit', used with divinatory meaning in 4Q186), אות ('sign', close to Greek *semeion*). Progressive revelation was necessary in order to improve the diagnostic skills of the Maskil, a field which was not well-represented in earlier apocalyptic literature. In other words, I suggest that Yahad members—or at least the Maskil—pursued new developments in the field of diagnostic astrology because they believed this was part of the mysteries revealed to them, which could be as useful as the study of Torah. After all, as noted above, astrology was part of the רז נהיה, which is promoted in the wisdom texts as a source of inspiration and the object of constant reflection. Indeed, an examination of the yahad's teachings reveals that in the same way that revelation led to novel torah rulings, the yahad also introduced novel insights to the astrological practice, using methods that are hardly attested elsewhere and should be considered as novelties of the Yahad. These novelties, described below, which appear in explicit sectarian contexts, must

[82] Translation from Qimron and Charlesworth, "Rule of the Community (1QS)", 41 with minor variations.
[83] See note 31 above.

have been justified by some sort of ideology about the role of astrology in unfolding the Divine wisdom.

Returning now to the above discussion about the role of astrology in the Yahad, new notions can be pointed out with the creative scientific character of the Yahad in mind. The discussion so far has highlighted the role of astrology in the wisdom texts as well as in the *Serekh*, a text which is strongly influenced by these traditions. However, a more tangible sense of the practice of astrology in the Yahad comes forth in the text 4Q186. This fragmentary text, dated to around the turn of the era, diagnoses a series of personal types by means of placing their moments of birth in a finely measured section of a zodiacal sign. The fine tuning extends beyond the assignment of one sign (=30°) for the person, as attested for example in birth omens from Mesopotamia, employing instead a more advanced location. In one extant example the birth is located within a section of the sign Taurus ברגל השור 'in the foot of Taurus'; in addition there seems to be a finer division of the sign into nine or more parts, divided between the 'house of light' and the 'house of darkness'. The fine division of the zodiacal sign was explained by Albani and others as indicating the 'ascendant' (Greek ὡροςκόπος), an astrological technique which focuses on the exact section of the zodiacal sign rising above the horizon at the time of birth.[84] Most interestingly for the present purposes, the ascendant could not have been computed before Hypsicles of Alexandria in the early second century BCE, and did not enter the astrological practice until later. The ascendant was thus a relatively new concept.[85] Its actual use in Greek horoscopes does not precede 62 BCE, according to the material currently known to us. While 4Q186 does not formally answer our definition of a horoscope, its use of the ascendant makes it one of the earliest pieces of

[84] This was first noted by Albani in 1993, conveniently approached in Albani, "Horoscopes," 305-309, and maintained by Popović, *Reading the Human Body*, 164-171, and Leicht, *Astrologoumena Judaica*, 24-28. A different interpretation was given by Francis Schmidt, "Astrologie juive ancienne: Essai d'interprétation de 4QCryptique (4Q186)" *RQ* 18 (1997): 125-141. Popović surveys earlier literature and supplies numerous new insights into the zodiacal astrology employed in this text. See also below.

[85] See Popović, *Reading the Human Body*, 125 and the rich bibliography cited there.

manuscript evidence for the ascendant in any language, Greek included, and certainly the earliest in Jewish literature. The Yahad sectaries thus not only subscribed to the antique science of their Enochic predecessors, but also sought to renew the techniques available to them.[86] In the present case, the motivation for improving the astrological techniques was the need for a better diagnostic of group members. Such an activity could not have taken place without being accommodated into the epistemological paradigm of the Yahad, either explicitly or implicitly.

A closer look at 4Q186 will supply further insights about the scientific discourse in the Yahad, and will also correspond to section (b) of the above discussion: translation and accommodation of earlier knowledge. The text is concealed from the lay reader by a strange mechanism of writing from left-to-right, while in addition some of the Hebrew letters are replaced with cryptic characters of sorts.[87] Therefore I see this text as a sectarian production. A characteristic quote from this scroll reads as follows (4Q186 1 ii 5-9):[88]

ושוקיו ארוכות ודקות ואצבעות רגליו
דקות וארוכות והואה מן העמוד השני
רוח לו בבית האור שש ושלוש בבית֯
הח֯ושך וזה הואה המולד אש⌝הואה ילוד עליו
ברגל השור ענו יהיה וזה בהמתו שור

> and his thighs are long and slender, and his toes are slender and long. And he is from the Second Column. He has six (parts) spirit in the house of light, and three in the house of darkness. And this is the nativity on which he is

[86] This conclusion stands in contrast to an old supposition by Michael Stone. Stone claimed that the *Book of Astronomy* used old fashioned Mesopotamian science as an act of resistance to the force of the new and contemporary Greek science: Stone, "Enoch, Aramaic Levi and Sectarian Origins." While this notion is possible with regard to the Enochic science, it is not valid with regard to the activity of the Yahad. Cp. also Popović, *Reading the Human Body*, 223.

[87] Popović, *Reading the Human Body*, 227-230.

[88] Text from Popović, *Reading the Human Body*, 29. The translation follows Popović with several corrections. Naturally not all the peculiarities of this text can be explained here. For "the Second Column" see below, note 101.

> born — in the foot of *Taurus.* He will be humble; and this is his animal: *Taurus.*

This text apparently creates an algorithm for casting a person's horoscope, using his bodily traits (thighs, toes) as a point of departure. Searching thoroughly for a similar procedure to that of 4Q186 in other divinatory literature, Popović came up with very meager evidence. Astrology and Physiognomy are often interrelated in ancient literature, but usually the logical sequence between them works in the opposite direction: one departs from the given birth data and deduces from them the predicted physiognomic features of the person. However, in 4Q186 the opposite is the case. The deduction of astrological data from bodily features is a very rare one, indeed unattested in the divinatory literature published or discussed until today.[89] I suggest that this extraordinary procedure was conceived and developed within the Yahad, based on earlier astrological and physiognomic traditions, due to the special diagnostic needs of the community. It is thus not the fruit of Greek or Babylonian knowledge, but rather a novel creation, somewhat awkward in terms of scientific argumentation, but one that was meant to answer a pressing need in community life.

The knowledge contained in 4Q186 is no longer purely 'theoretical' science, as in the *Book of Astronomy* for example, but rather an application of scientific theories in actual life. Covering a series of various human types, 4Q186 bears the form of a continuous treatise, not of mere sporadic statements within a wide theological context. The scant remains of this scroll suffice to show that it had originally contained a series of diagnosed persons or types of persons, forming as it were a treatise on personalities and their physiognomic traits.

b. Translation, accommodation, language

As I claimed in more detail elsewhere, much can be made from the choice to write 4Q186 in Hebrew, using a certain sort of encryption, all the more so since a similar case is apparent also in the scroll 4Q317 cryptA Phases of the Moon.[90] Both scrolls depend to some

[89] Popović, *Reading the Human Body,* 112-118.

[90] Ben-Dov, "Scientific Writings in Aramaic and Hebrew." In that article I discussed the possibility that 4Q186 constitutes a translation—or possibly

extent on Aramaic precedents, with a pronounced new thrust in the Hebrew scroll. In addition, both are encrypted, probably in order to render the new text more authoritative, or possibly to limit its circulation to the initiated.[91] Translation and encryption are mutually related: the encryption assigned ritual significance to the new text once it was presented in the holy tongue.

Scientific themes were one of the few encrypted genres in the Qumran library. The other genres are wisdom (4Q298), Halakhah (*Midrash Sepher Moshe*, 4Q249) and community organization (*Serekh ha-'Edah*, 4Q249[a-i]). Why were these themes chosen? This question is part of a large set of questions arising from the corpus of writings in esoteric script at Qumran. These questions were discussed only preliminarily and require a more comprehensive treatment, an urgent desideratum in our field.[92]

Not all science in the scrolls was ideological and sacred, however. A good case in point is the scroll 4Q318, which is in Aramaic and not encrypted. Precious little information accompanies this scroll: it displays neither a narrative framework as in the scientific treatises of apocalyptic literature, nor a religious-organizational context, as in the

translation and revision—of the Aramaic scroll 4Q561 Horoscope. This question relies upon the more fundamental question whether at all there are astrological elements in 4Q561, or rather it relates to physiognomy only. Popović denies the presence of astrological elements in 4Q561, seeing it as a purely physiognomic treatise: Popović, *Reading the Human Body*, 54-55. Puech (DJD 37: 304), however, detected an astrological element in the (partly or fully reconstructed) term (רוח 4Q561 3 9; 6 2-4), connecting it with the similar usage in 4Q186. With Popović I consider the presence of astrological elements in 4Q561 unlikely; see also his response in Mladen Popović, "The Emergence of Aramaic and Hebrew Scholarly Texts: Transmission and Translation of Alien Wisdom," in *The Dead Sea Scrolls: Transmission of Tradition and Production of* Texts (eds. S. Metso, H. Najman and E. Schuller; STDJ 92; Leiden: Brill, 2010), 100-104.

[91] Compare the practice of writing cryptic texts in Hieratic papyri from Egypt: Jacco Dieleman, *Priests, Tongues, and Rites: The London-Leiden Magical Manuscripts and Translation in Egyptian Ritual (100-300 CE)*, (Leiden: Brill 2005), 80-87.

[92] See Pfann, "Writings in Esoteric Script"; idem, *The Character of the Early Essene Movement in the Light of the Manuscripts Written in Esoteric Scripts from Qumran* (Dissertation, Hebrew University of Jerusalem, 2001); Thomas, *The Mysteries*. Alan Lenzi tried, to no avail, to explain why some Mesopotamian scholarly texts are marked as secret while others or not, but produced extremely helpful statistics and illuminating reasoning: Lenzi, *Secrecy*, 204-219.

sectarian astronomy, astrology and calendar texts. Instead, the remains of this scroll preserve an innocent anthology of astrological lists, divided between a *bronotologion* (omina based on the appearance of thunder in specific months) and a *selendromion* (a roster of the place of the moon within the signs of the zodiac).[93] Both extant parts of this scroll find parallels in Greek and Babylonian astrological literature. There is little, however, which would connect this scroll with sectarian practice. The fact that it is written in Aramaic and lacks any sectarian terminology—indeed any theological and cultic terminology at all—also speaks against its demarcation as a document of the Yahad. That is to say, it is certainly an intriguing piece of ancient science, and one cannot deny that it was found at Qumran; however, it does not correspond to the patterns employed by the scholars of the Yahad, as exemplified in other scientific themes that were submerged in the literary stream of the community.

In the above mentioned article I suggested that linguistic ideology played a significant part in the scientific efforts of the Yahad.[94] In the present context, as a wider discussion of scientific themes in a theological framework is employed, we may note that linguistic ideology played a dominant part in legitimating science among Jews also in the Medieval period. A good example comes from the treatise, *Keley Nehoshet* by the great medieval astronomer, astrologer and Bible exegete Abraham Ibn Ezra:[95]

[93] Jonas C. Greenfield, Michael Sokoloff, and David Pingree, "318. 4QZodiology and Brontology ar," in *Qumran Cave 4 XXVI. Cryptic Texts and Miscellanea, Part 1* (eds. S.J. Pfann, P.S. Alexander et al.; DJD 36; Oxford: Clarendon Press, 2000), 259-274; Albani, "Horoscopes," 296-301; Mark J. Geller, "New Documents from the Dead Sea: Babylonian Science in Aramaic," in *Boundaries of the Ancient Near Eastern World: a Tribute to Cyrus H. Gordon* (eds. M. Lubetski, C. Gottlieb and S. Keller; JSOTSup 273; Sheffield: Sheffield Academic Press, 1998), 224-229; Ben-Dov, *Head of all Years*, 256-257; Helen R. Jacobus, "4Q318: A Jewish Zodiac Calendar at Qumran?," in *The Dead Sea Scrolls: Texts and Context* (ed. C. Hempel; Leiden: Brill, 2010), 365-395.

[94] Ben-Dov, "Scientific Writings". Puech made a similar statement in *DJD 37*, 305.

[95] Translation follows Shlomo Sela, *Abraham Ibn Ezra and the Rise of Medieval Hebrew Science* (Brill's Series in Jewish Studies 32; Leiden: Brill, 2003), 105. For further examples and a penetrating analysis of linguistic ideology in scientific writings see Sela, ibid., 93-143.

> ... the holy tongue was the most comprehensive since it was the first among the languages of all nations. But, since the holy people were exiled from their holy land, they intermingled with the other nations, learnt their languages, and so forgot their own language and were only left with the books of the Prophets. But all those words which the Prophets had no need for in their works, do not appear at all in scripture. Therefore it is difficult to create new nouns, that is, to translate them from one language into another.

Like other medieval writers, Ibn Ezra is required to cope with the fact that the dominant science in his days was not Jewish, and that the Jews interested in science were obliged to base themselves on the achievements of gentiles. Moreover, the Hebrew language was ill equipped to deal with the scientific curriculum, lacking proper terminology. Ibn Ezra produced an interesting explanation for this situation: while the Hebrew language knew all the required scientific terminology because it was by definition a perfect language, and because the ancient Hebrews mastered all scientific knowledge, this knowledge perished in the exile. Ibn Ezra's task is thus not to invent a new Hebrew scientific vocabulary and a fresh discipline, but rather to reclaim what had been Jewish originally![96]

As the finds presented here show, linguistic ideology and the dominance of Hebrew played a part in the Yahad's ideology which is not too far apart from that of Ibn Ezra.[97] The Yahad sought to construct a Jewish scientific corpus in the Hebrew language, based on antecedents which were mostly Aramaic. Several Hebrew terms like *dwq* 'waning of the moon', *mwld* 'nativity', and *ʿmwd* 'column in a table'.[98] were invented or refashioned for this purpose. The translation of science joins forces with the special importance attributed to the Hebrew language in Qumran texts and related literature. While Aramaic has been the general language of Apocalypticism, the Yahad

[96] See in much detail Sela, *Abraham Ibn Ezra* .

[97] For the contents of this paragraph see in greater detail Ben-Dov, "Hebrew and Aramaic Writing."

[98] For the latter see below, note 101.

departed from this tradition, presenting itself as exceeding mere visionary literature.

The intention to supersede previous apocalyptic literature is apparent in one further element of the pattern in which earlier scientific writings were taken over by the scholars of the Yahad. While the 'I' of the speaker in the scientific passages of Enoch and Levi was the primordial hero himself, as part of the literary framework of the book, the speaker in Yahad scientific documents is anonymous, just like the authoritative voice in all other sectarian Dead Sea Scrolls.[99] There is no longer need of an external figure to legitimize science, as it is considered a legitimate branch of knowledge in and of itself. While science in the Enochic tradition served as part of the eschatological paradigm, within the Yahad it was more tangibly integrated into the sectarian texture of beliefs, and, more importantly, of practices. Astrology, which had been entirely avoided in the Book of Astronomy, was applied in the Yahad as a means for enhancing the diagnostic ability of the Maskil. In a similar way, while the Book of Astronomy was never explicit about the use of its teaching as a cultic calendar, this usage is widely expanded in the sectarian Mishmarot texts.

A central feature of the scientific activity in the Yahad must finally be pointed out: its relative lack of mathematical skills, precise measuring methods, and structured presentation models. As we understand today, Science is not only the 'content' in whichever form it appears; rather it is the combination of form and content, as the formal means chosen by the scientific writers play a central part in fashioning their research questions and the entire scientific worldview.[100] However relativistic we bring ourselves to be in our judgment of pre-modern science, the fact should be made clear that its practice in the Yahad did not lead to new methods, nor to new observations, greater precision or novel mechanistic models. Scientific theories were used on a schematic level and harnessed to the ideology

[99] Popović, "The Emergence," 91-93, 97-99. See F. García-Martínez, "Beyond the Sectarian Divide: The 'Voice of the Teacher' as an Authority-Conferring Strategy in Some Qumran Texts," in *The Dead Sea Scrolls: Transmission of Traditions and Production of Texts* (eds. S. Metso, H. Najman and E. Schuller; STDJ 92; Leiden: Brill, 2010), 227-244.

[100] See mainly Karine Chemla (ed.), *History of Science, History of Text* (Boston Studies in the Philosophy of Science 238; Dordrecht: Springer, 2004).

and practice of the community. The Yahad scholars' notion of making sense of the world did not consist of accurate measurements and sets of data arranged in neat tables.[101] Thus for example, the lunar theory propagated in the Qumran calendars is twice or thrice removed from the Babylonian origins of this discipline: the measurements are not realistic but rather schematic, the system for expressing fractions is far less accurate than the cuneiform demarcation, and the tabular presentation is replaced by continuous prose. The acculturation of science in the Yahad, theologically elaborate as it was, tended to be more schematic than its extra-Israelite ancestors and relied on less accurate means of expression.

An additional manifestation of science at Qumran was not discussed until now: the so-called Qumran Sundial. Admittedly it is much harder to make sense of this find within the present discussion as it does not speak with words but rather with the mute force of an enigmatic object. Found in Locus 45 of Khirbet Qumran, the stone disc is about 14 cm in diameter, marked with an elaborate system of concentric rings and gradual marks, and furnishes a place for a gnomon in its middle. In 1997 the disc was identified as a sundial, with the *editio princeps* also offering a preliminary explanation for its mode of use and the underlying astronomical theory.[102] While the technical aspect of their explanation is insufficient, significant advance was

[101] One reference to tables remains in the Qumran scientific texts: the term העמוד השני (4Q186 1 ii 6; 4 1; 6 2), which refers in my opinion to the second column of an astrological-astronomical table. The term עמוד is comparable to Greek σελίς, 'cross-beam in ceiling construction' but also 'column of writing in a papyrus roll' (LSJ). By pure chance, δεύτερον σελίδιν "the second column" is partly preserved in the astronomical papyrus PSI 1491 line 4: Alexander Jones, "Babylonian Lunar Theory in Roman Egypt: Two New Texts," in *Under One Sky, Astronomy and Mathematics in the Ancient Near East* (eds. J. Steele and A. Imhausen; AOAT 297; Münster: Ugarit Verlag, 2002), 169. For the history of the use of tables in scientific texts see M. Campbell-Kelly et al., *The History of Mathematical Tables: From Sumer to Spreadsheets* (Oxford and New-York: Oxford University Press, 2003).

[102] Uwe Glessmer and Matthias Albani, "An Astronomical Measuring Instrument from Qumran," in *The Provo International Conference on the Dead Sea Scrolls: Technological Innovations, New Texts, and Reformulated Issues* (eds. D.W. Parry and E. Ulrich; STDJ 30; Leiden: Brill, 1997), 407-442. Glessmer and Albani published several modifications of their initial explanation. They are surveyed in detail together with other studies in Jonathan Ben-Dov "The Qumran Dial: Artifact, Text, and Context," in *Qumran und die Archäologie* (ed. J. Frey, C. Claussen and N. Kessler; Tübingen: Mohr Siebeck, 2011), 211-237.

achieved recently in a publication by P. Tavardon.[103] The presence of this instrument at Qumran—a sundial that does not resemble any other sundial known from Antiquity—could support the claim for the existence of scientific and technological ability among the Yahad members.

Unfortunately, cogent analysis casts serious doubts on the adequacy of this object as an astronomical measuring device. On the one hand it would have been of little practical use as a sundial, while on the other hand simple and fully functional Roman-type sundials were easily available in Judea at the time. Furthermore, the astronomical skills which would have been required to operate the sundial—if it indeed worked as scholars suggest that it did—do not correspond to the interests of similar devices from the ancient world, and the crude level of production would damage the degree of expected accuracy. In addition, there is meager correspondence between the interests of the calendrical and astronomical texts from Qumran and the data collected by this particular object. Based on these and other difficulties I therefore concluded my article about this disc by expressing doubt whether it 1) was used as a scientific measuring instrument, and 2) how much it could be connected with the Yahad.

If indeed a scientific instrument in sectarian use, despite the incongruence with the calendrical texts, one may suggest that the disc was manufactured for the sake of experiments by a person or persons who were not interested in the convenience of contemporary technology but chose instead to tread a new path, crude as it may be. At the same time that person was keen on making sense of the natural world. Thus, as much as the device seems naïve in terms of the history of technology, it might have been significant for a person who viewed the world through idiosyncratic eyes. This scenario—feeble as it is—epitomizes the scientific activity in the Yahad: driven by a strong motivation to make sense of the world, and having inherited some scientific techniques from previous cosmological traditions, members of the Yahad were free to improvise on scientific themes in an effort to merge them in the religious-apocalyptic fabric of the community.

[103] Paul Tavardon, *Le disque de Qumrân* (CRB 75; Paris: Gabalda, 2010).

5. Conclusion

The present study sought to uncover the Ideals of science in the Dead Sea Scrolls community on the background of previous sapiential and apocalyptic thought. After presenting a case study from the integration of astrological and astronomical themes in Yahad literature, the paper aimed to evaluate the epistemological infrastructure to this kind of activity. Three categories were chosen to calibrate the scientific ideals in the scrolls. Although they are by no means exhaustive, they stand as good indicators for the type of scientific reflection that could be expected. They are: a notion of the origin of knowledge in the world; a justification for absorbing and reworking earlier scientific material; and an active integration of the scientific material in religious and social life.

A Long section discusses the epistemology of science in the Enochic tradition, analyzing the narratives from the *Book of Watchers* and the Book of Astronomy according to the hereby suggested tools. In this tradition, knowledge was dependant on a one-time revelation given to a primordial patriarch. It is legitimized by being threaded into a narrative about that patriarch. Being an ancient tradition, it perpetuates in later generations as students are commanded to contemplate the wisdom of Enoch. However, this epistemological framework does not encourage creative scientific work.

The Yahad, in contrast, engendered more productive conditions for scientific creativity. This stemmed mainly from the paradigm of revelation and the learning processes employed in the Yahad: learners were encouraged to seek renewed revelation in a less hierarchic environment, not only with regard to the study of Torah and prophets, but also with regard to the natural world. Departing from the knowledge transmitted in apocalyptic writings, and encouraged by the cosmological ideology of 4QInstruction, scholars in the Yahad found new paths in astronomy and astrology. Those new paths depended on precedents from the *koine* of the time, while molding them to fit the needs of the community. The result was a kind of science that may seem awkward in modern eyes but is motivated by the religious-social needs of the community. I support this view by an analysis of 4Q186, which I see as an application of the vague astrological statements in

4QInstruction and in *Serekh ha-Yahad*. This scholarly endeavor was, in turn, buttressed by a linguistic ideology which sought to anchor the new scholarly achievements in the Hebrew language, while at the same time authorizing some of the texts by means of the code we know as Cryptic A. The enigmatic (sun)dial from Qumran may also fit in this paradigm.

In contrast to the activity of Jewish scientists in the medieval period, which was not inherently Jewish but is better seen as general science practiced by Jews, the science in the Yahad was Jewish in a deeper way. Whoever expected to find in this article the magnificent scientific achievements by Jews two millennia before Einstein, should be slightly disillusioned. What the modern science values as scientific skill was not particularly strong at Qumran. The scrolls do display, however, a lively Jewish scientific discipline, which anticipated the questions of medieval Jewish scientists.

7. Networks of Scholars: The Transmission of Astronomical and Astrological Learning between Babylonians, Greeks and Jews

Mladen Popović

1. *Introduction*

What do we know about what ancient Jewish scholars knew about what Babylonian scholars knew?[1] In order to answer this question we can analyse the different scholarly texts at our disposal, Babylonian and Jewish, and look for similarities and differences. We can then explain which Babylonian elements were familiar to Jewish scholars and how they appropriated, used and reworked these. Such analyses usually work from specific Jewish texts and then look for Babylonian elements, retracing these in specific cuneiform texts.

The issue of tracking influences and cultural encounters between Babylonia and Jewish Palestine has another side to it, one not often put to the fore. Previous research on tracing influences of Babylonian learning in ancient Jewish texts has reflected insufficiently on the specific nature of such cultural encounters and the means of transmission. It has been assumed, tacitly or explicitly, that Jewish scholars had direct access to Babylonian centres of learning.[2] On the level of textual comparisons, it is evident that elements of Babylonian

[1] In this article I develop further some of the arguments in my "The Emergence of Aramaic and Hebrew Scholarly Texts: Transmission and Translation of Alien Wisdom," in *The Dead Sea Scrolls: Transmission of Traditions and Production of Texts* (ed. S. Metso, H. Najman, and E. Schuller; Leiden: Brill, 2010), 81–114.

[2] This relates to primarily indirect evidence in ancient sources that has been adduced to support such inferences; one can think of the portrayal of Daniel's position at the Neo-Babylonian court. See, for example, the discussion and references in Jonathan Ben-Dov, *Head of All Years: Astronomy and Calendars at Qumran in their Ancient Context* (Leiden: Brill, 2008), 266, 270-275. See also Henryk Drawnel, *The Aramaic Astronomical Book (4Q208-4Q211) from Qumran: Text, Commentary, and Translation* (Oxford: Oxford University Press, 2011), 51, 53, 70, 301, 304, and the contribution by James VanderKam in this volume.

origin appear in ancient Jewish texts; this is not disputed at all. However, the occurrence of textual similarities alone does not constitute enough evidence to argue for a direct connection. We at least need to further qualify the nature of that connection. Considering the social context of knowledge transmission, we must ask how and through whom ancient Jewish scholars got to know about some of the things that Babylonian scholars knew.

In this article I wish to focus mainly on two aspects of the social context of the transmission of astronomical learning in ancient Judaism. The first aspect is the differentiation between various levels of learning on the one hand and how this manifests itself in social relations between learned individuals on the other. The second aspect is that of ethnicity and cultural encounters with regard to learned knowledge.

I shall argue that the level of learning that we encounter in early Jewish sources differs starkly from contemporary developments in Babylonian and Greek astronomical science and points to a different trajectory of transmission. The transmission of Babylonian learning to Jewish scholars in Palestine was not a direct one. Furthermore, it went through different channels from the transmission of Babylonian astronomical science to Greek scholars. We should not assume as fact that Jewish scholars had direct access to Babylonian schools. We have no evidence for this. Due to the nature of the evidence at our disposal we cannot be very specific about the exact ways in which Babylonian scholarship reached Jewish Palestine sometime during the second half of the first millennium BCE Yet from a methodological point of view a number of considerations discussed in this article should be taken into account in future research on this issue.[3]

[3] Regarding the concept of science in Antiquity, I am not interested in any normative evaluations about whether the ancient learning we encounter in these texts should properly be called "science." Science is not detached from social reality. It is to a degree a historically defined activity conducted by people in different contexts. What counts as scientific knowledge may differ over time and place depending on context; see, e.g., David N. Livingstone, *Putting Science in Its Place: Geographies of Scientific Knowledge* (Chicago: The University of Chicago Press, 2003); Mladen Popović, *Reading the Human Body: Physiognomics and Astrology in the Dead Sea Scrolls and Hellenistic-Early Roman Period Judaism* (Leiden: Brill, 2007), 211-213; Eleanor Robson, "Empirical Scholarship in the Neo-Assyrian Court," in *The Empirical*

2. Babylonian and Jewish Astral Sciences

A brief survey of the Enochic astronomical material and some of the Dead Sea Scrolls from Qumran with regard to the Babylonian elements that they contain[4] provides some insight into the interests and level of knowledge of ancient Jewish scholars.[5] The point is to

Dimension of Ancient Near Eastern Studies / Die empirische Dimension altorientalischer Forschungen (ed. G.J. Selz and K. Wagensonner; Vienna: Lit, 2011), 603-629.

If need be, working definitions for ancient science can be given; see, e.g., Philip S. Alexander, "Enoch and the Beginnings of Jewish Interest in Natural Science," in this volume; Francesca Rochberg, *The Heavenly Writing: Divination, Horoscopy, and Astronomy in Mesopotamian Culture* (Cambridge: Cambridge University Press, 2004); Ben-Dov, *Head of All Years*; Rita Watson and Wayne Horowitz, *Writing Science before the Greeks: A Naturalistic Analysis of the Babylonian Astronomical Treatise Mul.Apin* (Leiden: Brill, 2011).

The choice of what count as scientific texts here has been made pragmatically and is determined by specific Jewish texts and their Babylonian "counterparts." To a certain extent, the demarcation between various kinds of texts is, of course, arbitrary. In this article I do not include, for example, magical texts, although elements that we might denote as magical also play a role in some of the texts under consideration here; see Popović, *Reading the Human Body*, 51–54, 234-237; idem, "Astrologische und magische Traditionen im antiken Judentum und die Texte vom Toten Meer," in *Qumran aktuell: Texte und Themen der Schriften vom Toten Meer* (ed. S. Beyerle and J. Frey; Biblisch-Theologische Studien 120; Neukirchen-Vluyn: Neukirchener, 2011), 111-136 (112-113, 127-131).

Furthermore, if we take into account the people that worked with these texts, demarcations are also to a certain extent arbitrary: there is evidence that different specialists in Mesopotamia worked on diverse and distinct disciplines, but there is also evidence that learned disciplines were not always nicely divided between different types of specialists. We need to look at specific historical contexts to determine whether demarcations were upheld and to what extent, and ask what this means for the dissemination and transmission of learned knowledge in that particular context. With regard to Jewish scholars or scientists in Hellenistic and early Roman Palestine, we have no comparative evidence about different kinds of scholars being responsible for different types of learning, although some such distinctions may be hypothesized; see the discussion and references in Popović, "The Emergence of Aramaic and Hebrew Scholarly Texts," 106–8.

[4] For the possibility that the numbers and fractions used in the *Aramaic Levi Document* reflect a Babylonian-type sexagesimal numeral system, see Henryk Drawnel, *An Aramaic Wisdom Text from Qumran: A New Interpretation of the Levi Document* (Leiden: Brill, 2004), 280-293. See also Ben-Dov, *Head of All Years*, 254; Popović, "The Emergence of Aramaic and Hebrew Scholarly Texts," 106–7.

[5] It is a moot point whether the scientific manuscripts from Qumran are representative for Jewish scholarship in general at the time or more typical for the movement behind the Dead Sea Scrolls. We know all ancient Jewish

discern some of the social and cultural aspects that may have determined the context of transmission of Babylonian elements of astral sciences in Jewish texts, and to consider this in comparison with the transmission of astral sciences within Babylonian culture and between Babylonia and the Greek world.

On the one hand, the Jewish texts attest to knowledge of some elements from Babylonian astronomy from the first half of the first millennium BCE (Enūma Anu Enlil and Mul.Apin), of the concept of the zodiac that was introduced in the fifth century BCE and of the non-mathematical Lunar Three scheme that was developed sometime later. On the other hand, these texts show an apparent ignorance with regard to sophisticated forms of mathematical astronomy that were developed in Seleucid Babylonia and transmitted to the Greek world.

The scholars behind the Jewish texts remain anonymous. In the Enochic corpus the authorial "I" was ascribed to Enoch, while the astronomical and astrological manuscripts from Qumran lack an authorial "I" altogether, as I emphasized elsewhere,[6] as do the calendric texts.[7] What Babylonian elements do we encounter in these texts? What changes and transformations did Babylonian elements go through as they were used in new Jewish-Palestinian texts and contexts?

2.1 Textual Comparisons

The earliest Jewish astronomical work is the Enochic *Astronomical Book* (extant after extensive redaction and modifications in 1 Enoch 72–82). Composed in Aramaic in the third century BCE or even earlier, the most complete text is preserved in Ethiopic translation. Four Aramaic copies (4Q208–211), containing only the *Astronomical Book*,

scientific literature, except the Enochic, only via the Qumran corpus. I assume that Jews outside this specific movement were also acquainted with this type of knowledge. The dates for some of the Aramaic *Astronomical Enoch* manuscripts predate the settlement at Qumran and indicate scholarly activity elsewhere and outside of that specific movement. Some of the other scholarly texts may likewise have circulated also outside of that movement; see Popović, *Reading the Human Body*, 8-11.

6 See Popović, "The Emergence of Aramaic and Hebrew Scholarly Texts," 87–93, 97, 99, 110 n. 72.

7 The temporal horizon of the Jewish texts is determined by the dates of the manuscripts from Qumran; these range broadly from the third century BCE to the first century CE.

have turned up at Qumran. The oldest fragments date to the late third or early second century BCE (4Q208). Scholars have pointed to a Mesopotamian background for some of the astronomical aspects of the Enochic *Astronomical Book*.[8] 1 Enoch 72 records the annual variation in the length of day and night-time. This variation is measured on a scale of eighteen, reflecting an M:m ratio of 2:1,[9] which results in a simple linear zigzag function rooted in Babylonian astronomy. Originally counting a 360-day year, as does Mul.Apin, for example, the *Astronomical Book* subsequently developed a 364-day year tradition.[10] For the so-called synchronistic calendar in the Aramaic fragments from Qumran (4Q208–209), the most recent suggestion is that it mainly deals with the duration of lunar visibility during night and day in a fashion similar to Tablet 14 of Enūma Anu Enlil.[11]

The astronomy in the Enochic *Astronomical Book* reaches back to older Mesopotamian examples from the first half of the first millennium BCE, such as Enūma Anu Enlil and Mul.Apin, but the *Astronomical Book* does not reflect developments that occurred in

[8] See, e.g. Otto Neugebauer, "The 'Astronomical' Chapters of the Ethiopic Book of Enoch (72 to 82)", in: *The Book of Enoch or 1 Enoch: A New English Edition* (M. Black; Leiden: Brill, 1985), 386–414 (387, 394-395); Matthias Albani, *Astronomie und Schöpfungsglaube: Untersuchungen zum astronomischen Henochbuch* (Neukirchen-Vluyn: Neukirchener, 1994), 155-172; Henryk Drawnel, "Moon Computation in the *Aramaic Astronomical Book*," *RQ* 23/89 (2007): 3–41; Ben Dov, *Head of All Years*; Drawnel, *The Aramaic Astronomical Book (4Q208-4Q211) from Qumran*. For the Mesopotamian background of other features as well, see, e.g., James C. VanderKam, *Enoch and the Growth of an Apocalyptic Tradition* (Washington, D.C.: Catholic Biblical Association of America, 1984); Helge S. Kvanvig, *Roots of Apocalyptic: The Mesopotamian Background of the Enoch Figure and the Son of Man* (Neukirchen-Vluyn: Neukirchener Verlag, 1988); idem, *Primeval History: Babylonian, Biblical and Enochic* (JSJSup 149; Leiden: Brill, 2011).

[9] 'M' stands for the maximum limit, while 'm' stands for the minimum limit. On the incorrectness of this ratio for Mesopotamia's latitude, see, e.g., Hermann Hunger and David Pingree, *Astral Sciences in Mesopotamia* (Leiden: Brill, 1999), 47.

[10] Ben-Dov, *Head of All Years*, 161-167, 182-183, 245-246 suggests that the Jewish 364-day year tradition actually goes back to an intercalation passage in Mul.Apin that implies the same number, but was never implemented in the actual astronomical models and soon after the seventh century BCE yielded to more accurate numbers. In Jewish astral science as we find it in the Qumran texts, however, the 364-day year model became the cornerstone of Enochic cosmological learning.

[11] Drawnel, *The Aramaic Astronomical Book (4Q208-4Q211) from Qumran*, 237–311.

Mesopotamian astronomy from the middle of the first millennium onward. These developments comprised the formulation of mathematical models that enable the prediction of the recurrence of certain astronomical phenomena and the introduction of the zodiac as a schematic, symbolic division of the ecliptic for computing and recording the planetary positions more exactly. The composition of the *Astronomical Book*[12] probably predates the even more advanced developments in Babylonian astronomy of the second century BCE, which were not incorporated into the text. The later copies of the Aramaic *Astronomical Book* from Qumran (4Q209–211 date to the second half and the end of the first century BCE) do not show evidence of such "updates." Moreover, the importance attributed to the Enochic type of astronomy is illustrated by the fact that the Aramaic *Astronomical Book* influenced other texts of the Jewish-Palestinian astronomical tradition, such as 4Q317 and 4Q503 and possibly 4Q334, Hebrew manuscripts dating from the second half of the second century BCE (4Q317 and 4Q503) and from the turn of the era (4Q334).[13]

There is also possible evidence for the use of more contemporary astronomical learning from Babylonia, although still in the form of non-mathematical astronomy. Scholars have interpreted some of the Qumran calendar texts in Hebrew (4Q320, 4Q321 and 4Q321a, which date to the end of the second century BCE until the second half of the first century BCE) to be based on elements of a Late-Babylonian lunar system (Persian-Hellenistic period), the Lunar Three scheme from non-mathematical astronomical texts, which were appropriated and modified by Jewish scholars to meet their own calendric needs.[14]

In addition to the Aramaic *Astronomical Book*, the texts influenced by it and some of the calendar texts from Qumran, there are two astrological manuscripts from Qumran to take into account. Both texts attest to knowledge of the zodiac and thus illustrate that they had

[12] On the distinctions between the different manuscripts and their relation to the identification and date of the *Astronomical Book* as a composition, see the discussion and references in Ben-Dov, *Head of All Years*, 69-118; Drawnel, *The Aramaic Astronomical Book (4Q208-4Q211) from Qumran*, 28-30, 39-53.

[13] See the discussion of these texts in Ben-Dov, *Head of All Years*.

[14] See Jonathan Ben-Dov and Wayne Horowitz, "The Babylonian Lunar Three in Calendrical Scrolls from Qumran," *ZA* 95 (2005): 104-120; Ben-Dov, *Head of All Years*, 197–244.

taken up one of the developments of Babylonian astronomy from the second half of the first millennium. However, whether this knowledge originated from Babylonia or from the Greek world is another matter.

Alongside Babylonian elements, these two texts appear to display Hellenistic features as well. This should not surprise us. During the Hellenistic period knowledge of Babylonian astrology and mathematical astronomy reached the Greek world, and having been transformed into Hellenistic astronomy and astrology it was taken elsewhere again. The Aramaic manuscript 4Q318 (copied around the turn of the era) consists of two parts. The first part (*selenodromion*) describes the synodic movement of the moon through the zodiac during twelve months of thirty days each, counting a 360-day year, as in Babylonian tradition. The second part (*brontologion*) has predictions for when it will thunder. This sort of text appears both in the Babylonian and Graeco-Roman astrological traditions.[15] The 360-day year scheme suggests a derivation from Babylonian tradition, but the zodiacal names in 4Q318 seem to point to Hellenistic origins.[16] The Hebrew manuscript 4Q186 (copied around the turn of the era) is a physiognomic-astrological catalogue combining different forms of learning (physiognomics,[17] astrology and possibly medicine and

[15] Popović, *Reading the Human Body*, 128.

[16] See, e.g., Jonas C. Greenfield et al., "4QZodiology and Brontology ar," in *Qumran Cave 4.XXVI: Cryptic Texts and Miscellanea, Part 1* (ed. P. Alexander et al.; DJD 36; Oxford: Clarendon, 2000), 259-274; Reimund Leicht, *Astrologumena Judaica: Untersuchungen zur Geschichte der astrologischen Literatur der Juden* (TSMJ 21; Tübingen: Mohr Siebeck, 2006), 23–24. See also Helen R. Jacobus, "4Q318: A Jewish Zodiac Calendar at Qumran?," in *The Dead Sea Scrolls: Texts and Context* (ed. C. Hempel; Leiden: Brill, 2010), 365-395.

[17] There is no evidence for astrology in the Aramaic text 4Q561; see Popović, "The Emergence of Aramaic and Hebrew Scholarly Texts," 102. 4Q561 (copied in either the first or the second half of the first century BCE) is a physiognomic catalogue that listed the physical descriptions of different types of people. There is some evidence that the text originally listed prognostics for each physiognomic type. Generally speaking, the physiognomic traditions of Babylonia and Greece are different. Babylonian physiognomics was principally a divinatory art that predicted people's future on the basis of their physical characteristics. Graeco-Roman physiognomics was by and large concerned with the discernment of people's characters, whereas the predictive function was minimal. The evidence for predictions in 4Q561 seems, therefore, to suggest Babylonian influence. But matters are not as clear-cut, because Babylonian physiognomics also seems to have been partially concerned with the

magic). The combination of different scientific disciplines within one text is probably already the case in Babylonian traditions predating the Hellenistic period.[18] The text's astrological framework, however, points decisively to a Hellenistic background, as the horoscope (מולד; *molad*) or ascendant (the point of the zodiac rising above the eastern horizon at the moment of birth) is of no importance at all in Babylonian horoscopy but is significant in Hellenistic horoscopy.[19]

Three observations can be drawn from the previous survey. First, elements from older types of Mesopotamian astronomy and astrology such as Enūma Anu Enlil and Mul.Apin from the first half of the first millennium BCE resurfaced in Aramaic in the early Enochic corpus sometime in the third century BCE[20] In this form, this type of astronomy continued to influence new compositions in Hebrew up until at least the second half of the second century BCE (4Q317 and 4Q503) and possibly later (4Q334). The evidence of new Hebrew compositions inspired by Enochic astronomy is important, as it demonstrates that the reception of older types of Babylonian astronomy was not limited to the early Enochic corpus. From a diachronic perspective we may observe that older types of Babylonian astronomy did not become obsolete after their initial reception in Jewish circles in the form of Enochic astronomy, but continued to be regarded relevant. This is demonstrated both by the later date of the *Astronomical Book* manuscripts from Qumran (4Q209–211) and by the new compositions in Hebrew that were inspired by it. This

discernment of character, and the case of Polemo of Laodicea (second century CE) demonstrates that Graeco-Roman traditions were also familiar with the predictive possibilities of physiognomics. See Popović, *Reading the Human Body*, 69–71, 111-112; Simon Swain (ed.), *Seeing the Face, Seeing the Soul: Polemon's Physiognomy from Classical Antiquity to Medieval Islam* (Oxford: Oxford University Press, 2007).

18 See, for example, the *Esoteric Babylonian Commentary* and *LBAT* 1593; Popović, *Reading the Human Body*, 112-114, 213-215.

19 Popović, *Reading the Human Body*, 123-125.

20 Maybe the composition should be dated to the fourth century already, although I see insufficient reason to assume a date for the *Astronomical Book* that much earlier than the manuscript evidence of 4Q208, especially since it is debatable whether the Aramaic Enoch manuscripts from Qumran should be equated with the *Astronomical Book* from Ethiopic Enoch; see the references in n. 12 above for a discussion of these issues.

preservation of older types of material may argue against continuing contact with Babylonia or at least may suggest rare contact.

Second, alongside the ongoing transmission of older elements of Babylonian astronomy from Enūma Anu Enlil and Mul.Apin, more recent features appear that post-date that type of astronomy but predate developments in the advanced Babylonian mathematical astronomy of the second century BCE. Calendric manuscripts from Qumran (4Q320, 4Q321 and 4Q321a) seem to be influenced by the Lunar Three scheme from Late-Babylonian non-mathematical astronomical texts. Astrological manuscripts from Qumran (4Q186 and 4Q318) demonstrate that knowledge of the zodiac, which was developed sometime in the fifth century BCE, found its way into Jewish society during this period as well. The manuscript copies are evidence of this development in the first century BCE at the latest. However, the mathematical type of astronomy is absent in all this.

Third, these ancient Jewish sources show a level of astronomical competency that was far less sophisticated than the advanced methods of contemporary astronomers in Babylonian and Greek societies. Starting in the Persian period but especially in the Seleucid period, Babylonian scholars made new and great advances in mathematical astronomy.[21] These advanced Babylonian astronomical methods were also transmitted to the Greek world. Due to the encounter with Babylonian astronomy and due to the work of Hipparchus and others such as Hypsicles, Hellenistic astronomy changed tremendously in this period.[22] Knowledge of certain elements of Babylonian astronomy among Greek scholars already occurred before Hipparchus, between the fifth and third centuries (Eudoxus, for example), but it is in the

[21] For an excellent overview of the developments in Babylonian astronomy and astrology, see Hunger and Pingree, *Astral Sciences in Mesopotamia*.

[22] See, e.g., Vittorio de Falco, Max Krause, and Otto Neugebauer, *Hypsikles: Die Aufgangszeiten der Gestirne* (AAWGPHK 3/62; Göttingen: Vandenhoeck & Ruprecht, 1966); Gerald J. Toomer, "Hipparchus and Babylonian Astronomy," in *A Scientific Humanist: Studies in Memory of Abraham Sachs* (ed. E. Leichty, M. deJong Ellis, and P. Gerardi; Philadelphia: The University Museum, 1988), 353-362; Alexander Jones, "The Adaptation of Babylonian Methods in Greek Numerical Astronomy," *Isis* 82 (1991): 441–453; idem, "Evidence for Babylonian Arithmetical Schemes in Greek Astronomy," in *Die Rolle der Astronomie in den Kulturen Mesopotamiens* (ed. H.D. Galter; Graz: GrazKult, 1993), 77–94; Rochberg, *Heavenly Writing*, 237-244.

second century, especially with Hipparchus it seems, "that we see a more complex and systematic exploitation of the resources of the Babylonian astronomers, including not only a whole array of period relations, but also extensive observations and, most notably, mathematical methods."[23]

The extant Jewish sources do not attest likewise to this influence from advanced mathematical astronomy from Babylonia. Aramaic Enochic astronomy remained non-mathematical. The earliest of the calendric manuscripts from Qumran (4Q320) dates to the end of the second century BCE and the youngest (4Q321) to the second half of the first century BCE While these manuscripts postdate the development of advanced astronomy in Babylonia and its transmission to the Greek world, they do not reflect those developments. The astrological manuscripts from Qumran attest knowledge of the zodiac, but this occurs in relatively straightforward lists. We do not find actual horoscopes, although even those need not presuppose the ability to execute complex observations and calculations. Such data may have been at hand in ephemerides and the like, but we have no evidence for this in Palestinian Judaism in this period. It appears that the concept of the zodiac found its way into Palestinian Jewish astral science as a finished product, so to say, without specialist knowledge of the mathematical-astronomical intricacies on which horoscopic astrology was based, and possibly also without access to supporting texts such as ephemerides—there is at least no need to suppose that these sorts of texts were available to Jewish scholars.[24]

2.2 Different Levels of Learning

Should we think of Enochic astronomy and its offshoots as survivals of types of astronomy that were already out-dated in Babylonia itself? What does "out-dated" mean? More advanced

[23] Toomer, "Hipparchus and Babylonian Astronomy," 353-354 (quote from p. 354).

[24] Popović, *Reading the Human Body*, 160-163. Previously I assumed that this absence of evidence does not imply evidence for the lack of ephemerides and almanacs, but I also emphasized that complex mathematical-astronomical knowledge was not a prerequisite for practising astrology. Now I also tend to emphasize that the absence of evidence should caution us not to tacitly assume that Jewish scholars would have had access to resources such as ephemerides and almanacs.

mathematical methods had been developed in Babylonian astronomy whilst Enochic astronomy flourished in Palestinian Judaism, but perhaps as a rule of thumb that kind of astronomy may still have had its worth. In that respect we should not unnecessarily devalue it. Furthermore, there is some evidence for the continued transmission of Enūma Anu Enlil and Mul.Apin in Babylonia during the Seleucid period, but we do not know whether these were actually used.[25] Even if there is some evidence for continued copying of Enūma Anu Enlil and Mul.Apin, it is evident that from a general perspective the character of astronomy differs strikingly between Babylonia and Jewish Palestine in the Hellenistic period. In Babylonia these texts were being copied alongside new and important astronomical advances, which are conspicuously absent from the extant Palestinian Jewish astronomical and astrological texts.

The extant texts thus show different levels of astronomical learning. What ramifications does this conclusion have for understanding the channels of transmission through which ancient Jewish scholars learned about Babylonian astral sciences? How did Jewish scholars in Palestine relate to the current advanced astronomical developments in Seleucid Babylonia? Did Jewish scholars consciously ignore these developments for some reason? Was the more recent and advanced Babylonian mathematical astronomy too difficult for them to appropriate? Or were they completely unaware of these developments because they were not connected in the same way to the same networks as those that transmitted Babylonian learning to the Greek world?

It is doubtful whether the apparent lack of knowledge of advanced Babylonian mathematical astronomy was due to a conscious decision motivated by certain theological considerations, as has been suggested.[26] A scientific-religious worldview characterized by a specific

[25] See Ulla Koch-Westenholz, *Mesopotamian Astrology: An Introduction to Babylonian and Assyrian Celestial Divination* (Copenhagen: Museum Tusculanum, 1995), 162; Rochberg, *Heavenly Writing*, 78. For text references, see Drawnel, *The Aramaic Astronomical Book (4Q208-4Q211) from Qumran*, 48 nn. 154-155. See also Petra D. Gesche, *Schulunterricht in Babylonien im ersten Jahrtausend v. Chr.* (Münster: Ugarit, 2001), 216.

[26] Cf. Ben-Dov, *Head of All Years*, 196. Cf. also Michael O. Wise, "Observations on New Calendrical Texts from Qumran," in *Thunder in Gemini and Other*

emphasis on heptadic-based numbers might explain the calendar texts from Qumran and the Enochic astronomical material to support that, but it does not provide a satisfactory explanation for the astrological material from Qumran. The concept of the zodiac does not support such a worldview.[27]

This begs the question: why were some concepts appropriated while others were not? There seems no reason to suppose that the authors or scribes behind the Enochic material and the Dead Sea Scrolls would have been intellectually less capable of dealing with complex mathematical-astronomical procedures from contemporary Babylonia. Instead, we should look for an explanation in terms of social relations and networks.

With due regard for the caveat that absence of evidence is not evidence of absence, in this case what is missing in the extant Jewish sources may provide us with a clue that, with regard to the astral sciences at that time, the channels of transmission between Babylonia and Jewish Palestine were probably different from the channels between Babylonia and the Greek world. When looking for Babylonian influences on ancient Jewish science it is not just a matter of listing the elements and looking for similarities and differences between texts. We cannot without further ado connect streams of traditions from Uruk, Sippar, Babylon and the like with Jerusalem, Qumran or any other place where Jewish scholars lived.

The different levels of learning should make us cautious of simply viewing the learned elites from different localities as interacting with each other on the same level. Mindful of Pierre Bourdieu's concept of cultural capital, we can appreciate the scholarly learning in the early Enochic and Qumran texts as a prized piece of knowledge signalling and confirming the status of those having access to and possessing it.[28]

Essays on the History, Language and Literature of Second Temple Palestine (Sheffield; Sheffield Academic Press, 1994), 222-239 (229).

[27] The astrological texts from Qumran represent learning that is different with regard to the heptadic-based model from Enochic astronomy and its offshoots and the calendric texts from Qumran. The notion of the zodiac is another crucial difference. The zodiac is absent in early Enochic astronomy.

[28] Mladen Popović, "Physiognomic Knowledge in Qumran and Babylonia: Form, Interdisciplinarity, and Secrecy," *DSD* 13 (2006): 150–176 (166-176); idem, *Reading the Human Body*, 81–83, 227-231.

But the Jewish scholarly elite may not have been of the same standing as the Babylonian scholarly elite, or the Greek one for that matter. In their respective societies Babylonian and Jewish scholars may have been members of the elite, but although stories about Abraham teaching Phoenicians and Egyptians or Daniel at the Babylonian court suggest otherwise, this need not imply that in real life the one recognized the other as equal or that there was direct contact between them, as is often assumed. Especially important here is to balance the two elements of having or gaining access and giving or allowing access as these materialize in social relationships between individuals.

3. Cultural Encounters and the Transmission of Astronomical Learning

Simplicius, the Late Antique commentator on Aristotle's works, wrote that, according to Porphyry, Aristotle had asked Callisthenes, who was in Babylon with Alexander the Great, to send astronomical observations from Babylon to Greece.[29] In *De caelo* Aristotle himself refers to observational data by the Egyptians and the Babylonians as the basis for much of their (i.e. the Greeks) evidence about particular luminaries.[30]

Regardless of whether such a request was really made by Aristotle, the anecdote illustrates how some in Late Antiquity imagined such cultural encounters and the transmission of knowledge between Babylon and Greece: reports were simply sent from Babylon directly to Greece by a Greek visitor of certain status. Unfortunately, the anecdote does not really provide us with much concrete information. The nature of the reports is not specified, nor their number: what kind of sources and how many? Nor is it clear what language they were in (a Late Antique writer would probably not have thought of Akkadian anymore):[31] did Simplicius imagine the reports

[29] FGrH 124 T 3.

[30] 292a. See Albert B. Bosworth, "Aristotle and Callisthenes," *Historia: Zeitschrift für alte Geschichte* 19 (1970): 407-413 (410-411). There is no reason to think that Aristotle made any such request to Callisthenes. More likely, Aristotle's reference in *De caelo* to observational data from the Babylonians set commentators thinking how the information could have been obtained, and one obvious hypothesis, it seems, was that Callisthenes supplied it, although this ignores the reference to the Egyptians.

[31] Cf., however, Markham J. Geller, "The Last Wedge," *ZA* 87 (1997): 43–95 (50). For an argument about the Greeks' lack of interest in learning others'

were in Greek? Furthermore, there is no explanation of how exactly Callisthenes obtained these sources or from whom. It was not as if astronomical reports were just lying around to be picked up by anyone. How did he know where to go? Did Callisthenes walk into a temple in Babylon, the Esangila, for example? Did he know individual Babylonian astronomers?[32] Such questions were probably not on Simplicius' mind, but from a socio-historical perspective on science we need to address some of these questions as to possession, accessibility and mobility in order to better understand the transmission of scholarly knowledge, in this case astronomical learning.

That Babylonian learning travelled far beyond its original geographic boundaries is undisputed. It is evident that Babylonian sciences such as astronomy and astrology were transmitted beyond the Mesopotamian cultural realm to the west, and to the east for that matter. We find Babylonian mathematical and astronomical formulas and calculations that turn up in roughly contemporary and later texts of other cultural realms. Modern scholars have shown that Babylonian astral sciences contributed to and influenced Greek astronomy and

languages, see Aage Westenholz, "The Graeco-Babyloniaca Once Again," *ZA* 97 (2007): 262–313 (275-278). See also Arnaldo Momigliano, *Alien Wisdom: The Limits of Hellenization* (Cambridge: Cambridge University Press, 1971); idem, "The Fault of the Greeks," *Daedalus* 104/2 (1975): 9–19 (15).

[32] Strabo, *Geogr.* 16.1.6, refers to the Orchenoi and the Borsippenoi as classes of Chaldean astronomers. In the same passage, Strabo also refers to individual Chaldean astronomers: Kidenas, Naburianus, Sudines and Seleucus of Seleuceia. For a brief discussion of individual astronomers known by name to Graeco-Roman authors as Chaldean astronomers, see Jones, "Evidence for Babylonian Arithmetical Schemes," 88–89; Geller, "The Last Wedge," 49; Tom Boiy, *Late Achaemenid and Hellenistic Babylon* (Leuven: Peeters, 2004), 297 n. 24. See also below in section 3.3 for an inscription from Larissa in Thessaly that refers to a certain Antipater from Hierapolis as a Chaldean astronomer. Of course, there are numerous references to the "Chaldeans" and their skills in Greek and Latin sources, but they function more often than not as stock characters for authors to talk about astrologers or their skills. Not much specific information as to the transmission of their learning or in what respect it was of Babylonian origin can be gleaned from these sources. The references to the expulsion of "Chaldeans" from Rome indicate that they were regarded as undesirable for various reasons at various times, but these references hardly give us specific data as to the concrete people and circumstances involved.

astrology.[33] Ancient sources also acknowledge such Babylonian origins,[34] and we have examples of Jewish texts that put Abraham on a par with scholars from Egypt, Phoenicia and Babylon.[35]

[33] See, for example, Otto Neugebauer, *A History of Ancient Mathematical Astronomy* (Berlin: Springer, 1975); Jones, "The Adaptation of Babylonian Methods"; idem, "Evidence for Babylonian Arithmetical Schemes"; idem, *Astronomical Papyri from Oxyrhynchus (P. Oxy. 4133-4300a)* (2 vols.; Philadelphia: American Philosophical Society, 1999); David Pingree, *From Astral Omens to Astrology: From Babylon to Bīkāner* (Rome: Istituto Italiano per l'Africa e l'Oriente, 1997); idem, "Legacies in Astronomy and Celestial Omens," in *The Legacy of Mesopotamia* (ed. S. Dalley et al.; Oxford: Oxford University Press, 1998), 125-137; Rochberg, *Heavenly Writing*; eadem, *In the Path of the Moon: Babylonian Celestial Divination and Its Legacy* (Leiden: Brill, 2010); John M. Steele, "Visual Aspects of the Transmission of Babylonian Astronomy and its Reception into Greek Astronomy," *Annals of Science* 68 (2011): 453-465.

[34] For example, Ptolemy of Alexandria, *Almagest* 3.7, refers to records of ancient observations from the beginning of the reign of the Babylonian king Nabonassar (747–734 BCE) that had been preserved down to his own time, and to which Ptolemy claimed to have had access, although no further information about them is provided, nor how Ptolemy accessed them and in what language. For a discussion of these issues, see, e.g., Jones, "The Adaptation of Babylonian Methods," 442-443; John M. Steele, "Applied Historical Astronomy: An Historical Perspective," *Journal for the History of Astronomy* 35 (2004): 337-355 (338-346). See also Christopher B.F. Walker, "Achaemenid Chronology and the Babylonian Sources," in *Mesopotamia and Iran in the Persian Period: Conquest and Imperialism 539-331 BC* (ed. J. Curtis; London: British Museum, 1997), 17–25; Steele, "Visual Aspects."

[35] Pseudo-Eupolemus tells of how Abraham taught astronomy and other sciences to the Egyptians when he dwelt with the Egyptian priests in Heliopolis (apud Eusebius, *Praep. ev.* 9.17.2–9). Presenting astral learning also as revealed knowledge (Methuselah learned everything through the angels of God), Pseudo-Eupolemus shares a *topos* with Babylonian sources. See, e.g., Wilfred G. Lambert, "A Catalogue of Texts and Authors," *JCS* 16 (1962): 59–77; Beate Pongratz-Leisten, *Herrschaftswissen in Mesopotamien: Formen der Kommunikation zwischen Gott und König im 2. und 1. Jahrtausend v. Chr.* (Helsinki: The Neo-Assyrian Text Corpus Project, 1999), 293-295; Rochberg, *Heavenly Writing*, 181-185, 215. The *Genesis Apocryphon* from Qumran Cave 1 also emphasizes Abraham's wisdom and learning, pointing out that the pharaoh's officials visited him to seek out his wisdom, as well as his wife. The *Genesis Apocryphon* possibly also refers more specifically to Abraham's astronomical wisdom, as he is said to have read to the Egyptian officials from the book of the words of Enoch (1QapGen ar 19:24–25). Josephus remarks that astronomical learning derived from the Chaldeans (*Ant.* 1.168). Philo admits as much, but explains Abraham's emigration from Ur allegorically as a rejection of astrology (*Abraham* 68–84; *Migration* 176–187; cf. also *Heir* 96–99; Josephus, *Ant.* 1.155–157). Likewise, Jub 12:16–18 has Abraham observing the stars before leaving Ur and his astrological interests behind. Some of these passages reflect an ancient discourse on the origins of astronomy/astrology, putting Abraham, via Enoch, forward as a Jewish

The views of Simplicius and, for example, Pseudo-Eupolemus,[36] are rather clear-cut: scientific data were transmitted to others in written form and scholarly knowledge was exchanged in direct interaction between learned men. In general, both aspects, written and oral communication, no doubt shaped ancient contexts of transmission, but in specific cases we probably have to differentiate: scientific data were not just sent around indiscriminately and not all learned men would, as a rule, have interacted with all learned men from other cultural realms, this being dependent on various factors.

Insights from Social Network Analysis theory may help to conceptualize specific conditions for the transmission of astronomical and astrological knowledge.[37] I will not provide a systematic consideration of all forms of connectivity that may have played a role,

culture hero. See brief discussion and references in Popović, *Reading the Human Body*, 225-226. The discourse on the origins of astronomy/astrology was in itself part of the wider debate about the priority of cultures and about which culture was dependent on which. See, e.g., Geert De Breucker, "Berossos between Tradition and Innovation," in *The Oxford Handbook of Cuneiform Culture* (ed. K. Radner and E. Robson; Oxford: Oxford University Press, 2011), 637-657 (650).

[36] See n. 35 above.

[37] The literature on Social Network Analysis theory is enormous, and it mostly concentrates on the study of the present, using statistical data. However, Social Network Analysis theory has also begun to be used for earlier periods of history. See, e.g., Peter S. Bearman, *Relations into Rhetorics: Local Elite Social Structure in Norfolk 1540–1640* (Brunswick: Rutgers University Press, 1993); John F. Padgett and Christopher K. Ansell, "Robust Action and the Rise of the Medici 1400–1434," *American Journal of Sociology* 98/6 (1993): 1259–1319; Charles Wetherell, Andrejs Plakans, and Barry Wellman, "Social Networks, Kinship and Community in Eastern Europe," *Journal of Interdisciplinary History* 24/4 (1994): 639-663; Barry Wellman and Charles Wetherell, "Social Network Analysis of Historical Communities: Some Questions from the Present for the Past," *History of the Family* 1 (1996): 97–121; B.H. Erickson, "Social Networks and History: A Review Essay," *Historical Methods* 30/3 (1997): 149-157; Paul D. McLean, *The Art of the Network: Strategic Interaction and Patronage in Renaissance Florence* (Durham: Duke University Press, 2007). In 2007 a special issue of *Mediterranean Historical Review* was devoted to Social Network Analysis. For more on Social Network Analysis theory and its application to the transmission of religious ideas and cults, see Anna Collar, "Network Theory and Religious Innovation," *MHR* 22 (2007): 149-162, and John Ma, "Peer Polity Interaction in the Hellenistic Age," *Past and Present* 180 (2003): 9–39. Although less on Social Network Analysis theory as such, see also Claudia Moatti, "Translation, Migration, and Communication in the Roman Empire: Three Aspects of Movement in History," *Classical Antiquity* 25 (2006): 109-140.

but will use Social Network Analysis theory as a heuristic tool to ask certain specific socio-cultural questions. The benefit of this approach is that it focuses our attention on the social context of the transmission of astronomical knowledge, as this is determined by social relationships, and with a special emphasis on networks. The question is what kinds of social networks were involved. People transmit information and knowledge via different mediums in specific contexts—defined by, for example, locality, social status, gender, age, kinship, nationality and ethnicity. In order to understand the context, i.e. the circumstances of these interactions and of the transmission of knowledge, we must focus on concrete localities, channels and agents.

In this section I wish to address three issues regarding the transmission of Babylonian astronomy and astrology: the involvement of non-Babylonians in Babylonian science, the role of the Aramaic language and the transmission to the Greek world specifically.

3.1 The Involvement of Non-Babylonians in Babylonian Science

With regard to ethno-linguistic and cultural distinctions, modern anthropological and cultural studies have called essentialist distinctions into doubt: what makes a certain type of learning "Egyptian" or "Babylonian," and how does that relate to concrete people, be they Egyptian, Babylonian, Greek or Jewish?[38] There is somewhat of a paradox in some recent analyses in that such terms are problematized as less useful heuristic concepts whereas our taxonomic interests call for their use. We wish to categorize and classify our data into neat and separate boxes, and at the same time wish to acknowledge that the boundaries between these conceptual boxes are often, in reality, fluid and fuzzy.

Nonetheless, that boundaries are fluid, fuzzy or that they can be crossed does not imply they do not exist; boundaries are not completely ephemeral, and sometimes they are very real. While ethno-linguistic and cultural borders can be crossed, they also function to create a persistent sense of difference.[39] Specific social and

[38] See, e.g., Annette Y. Reed, "Was there Science in Ancient Judaism? Historical and Cross-Cultural Reflections on 'Religion' and 'Science,'" *SR* 36 (2007): 461-495 (467), as well as her contribution in this volume.

[39] There is an enormous amount of scholarly discussion on these issues. For interesting insights in this discussion, from a different field of study than

cultural contexts will have determined when and how boundaries would have been more fluid or stricter.

We have evidence from the Neo-Assyrian period of non-Babylonians and non-Assyrians engaged in royal scholarly networks.[40] The late Assyrian empire had become bilingual and bicultural, with Aramaic becoming the international vernacular.[41] At the same time, there is evidence for adversarial reactions among the Assyrian ruling classes to the rising importance of Aramaic. For example, to a request by the scribe Sin-iddina of Ur to reply in Aramaic, the king answers that the scribe should rather write to him in Akkadian.[42] This example works both ways. On the one hand, it shows that ethno-linguistic boundaries were not strict: letters to the Assyrian king could be written in Aramaic. On the other hand, it demonstrates that such boundaries were not completely ephemeral. The sender of the letter may have asked for too much by requesting the king's answer also be in Aramaic. The king retorts by raising the boundary and emphasizing its importance: he creates an ethno-linguistic sense of difference between the scribe and himself. While ethno-linguistic and cultural boundaries can be crossed, they are also maintained. However, in the Neo-Assyrian period such boundaries do not seem to have prohibited the accessibility and dissemination of Babylonian science.

If we look at evidence from the Neo- and Late-Babylonian periods, it seems that the Babylonian urban elite had stricter limitations for entry into the scholarly elite than the Assyrians.[43] There is, however, some evidence for scribes bearing non-Akkadian names who copied

Jewish or cuneiform, see, e.g., David W. Anthony, *The Horse, the Wheel, and Language: How Bronze-Age Riders from the Eurasian Steppes Shaped the Modern World* (Princeton: Princeton University Press, 2007).

[40] See, e.g., Karen Radner, "The Assyrian King and his Scholars: The Syro-Anatolian and the Egyptian Schools," in *Of God(s), Trees, Kings, and Scholars: Neo-Assyrian and Related Studies in Honour of Simo Parpola* (ed. M. Luukko, S. Svärd, and R. Mattila; Helsinki: Finnish Oriental Society, 2009), 221-238.

[41] See, e.g., Paul-Alain Beaulieu, "Official and Vernacular Languages: The Shifting Sands of Imperial and Cultural Identities in First-Millennium B.C. Mesopotamia," in *Margins of Writing, Origins of Cultures* (2nd ed.; ed. S.L. Sanders; Chicago: The Oriental Institute, 2007), 191–220 (192-196).

[42] See the references in Beaulieu, "Official and Vernacular Languages," 193. See also Westenholz, "The Graeco-Babyloniaca Once Again," 293.

literary, scholarly texts: Šemaa, son of Adirum is thus far the only example, the extant sources showing that all Babylonian scribes with non-Akkadian names wrote documentary, legal texts.[44] Ethno-linguistic boundaries may not have been absolute, but some form of boundary maintenance with regard to literary and scholarly texts as distinct from documentary, legal texts did seem to be in effect.

Through cuneiform culture the Babylonian urban elite is said to have expressed a high degree of self-consciousness. For example, a cuneiform text from Hellenistic Uruk shows that the Aramaeans were still considered a separate ethno-linguistic group by some Babylonians;[45] the reference in the late Seleucid list from Uruk of kings and scholars to Esarhaddon's counsellor Aba-Enlil-dari as the one "whom the Aramaeans call Aḫuqar" shows that the story of Aḫiqar was known but seen as part of "popular" Aramaic culture rather than cuneiform elite culture.[46] The impression gained from cuneiform sources is of Late Babylonia as an imagined community of urban elites who retreated into the imaginary space provided by the temples and the schools, with cuneiform itself being the main distinguishing characteristic of this community. The Babylonian urban elites constructed a cultural identity for themselves, one that became more and more detached from the ethno-linguistic, cultural and political realities of Babylonia in the Hellenistic and Parthian periods.[47] At the

[43] Ran Zadok, "The Representation of Foreigners in Neo- and Late-Babylonian Legal Documents (Eighth through Second Centuries B.C.E.)", in *Judah and the Judeans in the Neo-Babylonian Period* (ed. O. Lipschits and J. Blenkinsopp; Winona Lake: Eisenbrauns, 2003), 471–589 (483-484); Beaulieu, "Official and Vernacular Languages," 212-213.

[44] Popović, "The Emergence of Aramaic and Hebrew Scholarly Texts," 110-114.

[45] Beaulieu, "Official and Vernacular Languages," 194-195.

[46] Cf. Westenholz, "The Graeco-Babyloniaca Once Again," 308. See also Beaulieu, "Official and Vernacular Languages," 194-195.

[47] Cf. Beaulieu, "Official and Vernacular Languages," 197, 213; De Breucker, "Berossos between Tradition and Innovation," 638-639; Philippe Clancier, "Cuneiform Culture's Last Guardians: The Old Urban Notability of Hellenistic Uruk," in *The Oxford Handbook of Cuneiform Culture*, 752-773 (756). See also David Brown, "Increasingly Redundant: The Growing Obsolescence of the Cuneiform Script in Babylonia from 539 BC," in *The Disappearance of Writing Systems: Perspectives on Literacy and Communication* (ed. J. Baines, J. Bennet, and S. Houston; London: Equinox, 2008), 73–101, who emphasizes the connection between the production and transmission of astronomical and

same time, evidence from Hellenistic Uruk seems to indicate that they cultivated strong ties with the Greek elite and the Seleucid rulers that ensured their small community thrived.[48] This shows that despite a cultivated identity that seems detached from real life, Babylonian urban elites were also able to relate to changing ethno-linguistic, cultural and political realities and to do so to their own advantage.

Changed historical circumstances between the late fifth and late fourth centuries and the second century BCE no doubt influenced the possibilities presented to and the choices made by those specialized in astronomy and astrology in the cities of Babylonia, determining also the mobility, accessibility and dissemination of that learned knowledge. For example, recent research on the collection owned by two separate families of *mašmaššu*'s in Uruk in the late fifth and late fourth centuries BCE suggests a tight social network of scholarly families.[49] The colophons suggest that this scholarly network operated on a limited geographical scale: scholars, students and their writings seem to have rarely travelled beyond Uruk or across professional divides.[50] If we consider these two families from the perspective of Social Network Analysis theory, such a limited network that consists of strong ties—kinship ties being the clearest example of strong ties—seems less conducive to bridge social boundaries and cross network distance between different ethno-linguistic groups.[51]

astrological texts and the survival of cuneiform until around the turn of the common era, and Jerrold S. Cooper, "Redundancy Reconsidered: Reflections on David Brown's Thesis," in *The Disappearance of Writing Systems*, 103–108.

[48] Clancier, "Cuneiform Culture's Last Guardians," 756-762.

[49] See Eleanor Robson, "The Production and Dissemination of Scholarly Knowledge," in *The Oxford Handbook of Cuneiform Culture*, 557-576; eadem, "Tracing Networks of Cuneiform Scholarship with Oracc, GKAB and Google Earth," in *Archaeologies of Text: Archaeology, Technology and Ethics* (ed. M. Rutz and M. Kersel; Oxford: Oxbow, forthcoming).

[50] The impression from the Neo-Assyrian period is that a lot of movement of both tablets and scholars went on. See e.g. Lorenzo Verderame, "La formazione dell'esperto (*ummânu*) nel periodo neo-assiro," *Historiae* 5 (2008): 51–67. The colophons on the cuneiform tablets from the late fifth- and late fourth-centuries BCE collection from Uruk often testify to being copies of originals that came from elsewhere, but we do not know how far back these originals were in the chain of transmission, see Robson, "The Production and Dissemination of Scholarly Knowledge," 566.

[51] Cf. Collar, "Network Theory and Religious Innovation," 151.

From a diachronic perspective, there is also evidence for non-familial apprenticeship in second-century BCE Seleucid Uruk. Although the social network there was tight, as over six generations just four scholarly families collaborated in the training of their sons, patterns of non-familial apprenticeship can be identified.[52] A limited network that consisted of strong ties (kinship) was thus able to bridge social boundaries and transmit learning to non-family. It is, however, not readily apparent whether these non-familial apprentices may be indicative as well for the crossing of ethno-linguistic and cultural boundaries, and if so, which ones exactly. Perhaps individual members from social networks such as these and others were able to cross much greater network distances: Uruk may have been one of the sources from which Babylonian astronomy made its way to the Greek world.[53] One can think of a scenario in which individual scholars from Uruk moved away from the city, taking their scientific knowledge with them. Historical and modern analogies suggest that this would have been mostly young men.[54] Perhaps such a move in the second century BCE, if it occurred at all, was due to economic reasons: private income generated by astronomical and astrological knowledge[55] may not have been sufficient, and in response some scribes took their chances and left.

In terms of ethno-linguistic and cultural boundaries the Neo- and Late-Babylonian evidence demonstrates complex and multi-layered contexts. On the one hand, ethno-linguistic distinctions were made and cuneiform culture was upheld to emphasize Babylonian identity. On the other hand, various sources show a multi-ethnic picture that seems to call into doubt the dominance of cuneiform culture in Late

[52] Robson, "The Production and Dissemination of Scholarly Knowledge," 565. See also Mathieu Ossendrijver, "Exzellente Netzwerke: Die Astronomen von Uruk," in *The Empirical Dimension of Ancient Near Eastern Studies*, 631-644.

[53] Cf. Jones, "Evidence for Babylonian Arithmetical Schemes," 88. See also below in section 3.3.

[54] See also Collar, "Network Theory and Religious Innovation," 156.

[55] Brown, "Increasingly Redundant," 89; Cooper, "Redundancy Reconsidered," 104; Eleanor Robson, "Reading the Libraries of Assyria and Babylonia," in *Ancient Libraries* (ed. J. Köning, K. Oikonomopoulou, and G. Woolf; Cambridge: Cambridge University Press, 2013).

Babylonia. This picture highlights the difficulty of distinguishing neatly between what is Babylonian and what is non-Babylonian.[56]

These two different aspects have direct bearing on the contexts of transmission of astronomical learning, and their complexity is well illustrated by the issue of Aramaic as a medium for the transmission of Babylonian astronomical and astrological learning.

3.2 Aramaic as a Medium for Transmission of Astronomical Learning

There is no evidence for the translation of complex astronomical cuneiform texts into Aramaic.[57] Aramaic *Astronomical Enoch* from Qumran does not even come remotely near to the advanced forms of cuneiform astronomy and it hardly can stand exemplary for a body of Babylonian Aramaic scholarship in the first millennium BCE

Apparently, the Babylonian urban elites kept to traditional cuneiform learning as the sole official culture of Babylonia. At the same time, Aramaic became the dominant vernacular. Should we then assume that there must have been a Babylonian literature and science written in Aramaic, now lost because it was written on perishable material? Or should we assume that Aramaic did not become the dominant language to express Babylonian high culture but remained mostly a language of communication and administration, and, moreover, that it is questionable whether any significant corpus of cuneiform was ever translated into Aramaic?[58]

[56] See, e.g., Amélie Kuhrt and Susan Sherwin-White (eds.), *Hellenism in the East: The Interaction of Greek and Non-Greek Civilizations from Syria to Central Asia after Alexander* (London: Duckworth, 1987); Ernie Haerinck, "Babylonia under Achaemenid Rule," in *Mesopotamia and Iran*, 26–34 (27); Robartus J. van der Spek, "Multi-Ethnicity and Ethnic Segregation in Hellenistic Babylon," in *Ethnic Constructs in Antiquity: The Role of Power and Tradition* (ed. T. Derks and N. Roymans; Amsterdam: Amsterdam University Press, 2009), 101-115.

[57] Contra Drawnel, *The Aramaic Astronomical Book (4Q208-4Q211) from Qumran*, 301, 304; see section 4 below.

[58] For the different positions in this debate, see, e.g., Beaulieu, "Official and Vernacular Languages," 198, 201, 212; De Breucker, "Berossos between Tradition and Innovation," 638; Clancier, "Cuneiform Culture's Last Guardians," 756. See also Sheldon Pollock, "Power and Culture Beyond Ideology and Identity," in *Margins of Writing*, 283-293 (esp. 285-286). For traces of first-millennium BCE Aramaic literature, see also Tawny L. Holm, "The Sheikh Faḍl Inscription in Its Literary and Historical Context," *Aramaic Studies* 5 (2007): 193–224 (220-224).

The Neo- and Late-Babylonian evidence presents us with different scenarios from the Neo-Assyrian evidence. We do not have a significant body of sources in Aramaic from first-millennium Babylonia or a clear and significant body of references from first-millennium Babylonia to the use of Aramaic and to the involvement of non-Babylonians in Babylonian science, as we do have regarding the last aspect for Assyria in the Neo-Assyrian period. Were ethno-linguistic and cultural boundaries in Babylonia more strictly maintained with regard to scholarly learning and did this inhibit the accessibility and dissemination of Babylonian science, or is it just a matter of evidence gone missing?

In addition to the Babylonian priests and scholars—the old urban notability—there was another group of scribes, the *sepīru* scribes. These did not belong to the old Babylonian nobility but were recognized as important persons who were well integrated in the temple organization.[59] They appear to have been Aramaic speakers who had become assimilated into Babylonian culture.[60] Although there is thus far no evidence that these scribes worked on literary or scholarly texts,[61] the question is how strictly boundaries were maintained. The *sepīru* scribes seem to have been well integrated into the temple organization, and the temple was *the* place of learning in Babylonia. The exact relations, however, between the Babylonian scholarly elite and Aramaic *sepīru* scribes merit further investigation. It might be that in the future evidence will turn up that *sepīru* scribes had access to cuneiform sources or that astronomical principles were divulged to them.[62]

We might also entertain the possibility that individual Jews, being part of a general Aramaic milieu, were among such Aramaic *sepīru*

[59] Philippe Clancier, "Les scribes sur parchmin du temple d'Anu," *RA* 99 (2005): 85–104 (93–98); idem, "Cuneiform Culture's Last Guardians," 764-765.
[60] Beaulieu, "Official and Vernacular Languages," 198.
[61] See the discussion in Popović, "The Emergence of Aramaic and Hebrew Scholarly Texts," 110-114.
[62] Interestingly, there are a couple of references to texts having been copied from scrolls—typically the material with which *sepīru* scribes are associated —but it is debated whether these references are to writings in Aramaic script. See, e.g., Clancier, "Les scribes sur parchmin," 90 n. 23; Westenholz, "The Graeco-Babyloniaca Once Again," 278-280; Cooper, "Redundancy Reconsidered," 106 n. 5.

scribes—perhaps having taken up Babylonian names even, making it more difficult for us to identify them as Jewish and thus also calling into question the aptness of such neat taxonomic distinctions. Be that as it may, no concrete evidence has as of yet turned up for Jews being Aramaic *sepīru* scribes. In the wake of the Babylonian conquest of Judah and Jerusalem in 586 BCE a Jewish Diaspora formed in Babylonia. Jews were fully integrated in economic everyday life there, as "is evidenced by their participation in very ordinary economic transactions in which they are recorded as the creditors and debtors in a variety of loan documents and receipts."[63] There is also enough evidence for Babylonian Jewish villagers or merchants in Neo- and Late-Babylonian documents, but there is no evidence for Jews accessing scholarly tablet collections directly.[64]

It is important to consider the role of the Aramaic language as a medium for the transmission of Babylonian astronomical and astrological material to the Jewish world. The westward transmission of this body of learned knowledge originating in cuneiform culture may have occurred through Aramaic sources, which is how Jewish scholars, being part of a general Aramaic milieu, may have encountered it.[65] The astronomical Enoch manuscripts from Qumran, for example, may be evidence for this role of Aramaic. However, whether it was through direct access to cuneiform texts, direct contact with Babylonian scholars in Mesopotamia or elsewhere, or indirectly via access to a—more vague—continuous tradition is another matter.

In order to better understand the possible role of Aramaic as a medium for the transmission of Babylonian astronomical material and also how some of this material made its way to Aramaic texts such as the astronomical Enoch manuscripts from Qumran, it is instructive to

[63] Laurie Pearce, "New Evidence for Judeans in Babylonia," in *Judah and the Judeans in the Persian Period* (ed. O. Lipschits and M. Oeming; Winona Lake: Eisenbrauns, 2006), 399–411 (402). See also F. Rachel Magdalene and Cornelia Wunsch, "Slavery between Judah and Babylon: The Exilic Experience," in *Slaves and Households in the Near East* (ed. L. Culbertson; Chicago: The Oriental Institute, 2011), 113-134.

[64] In general, the Babylonian temple archives indicate only a very small percentage of foreigners. See Zadok, "The Representation of Foreigners," 482-484.

[65] See discussion and references in Popović, "The Emergence of Aramaic and Hebrew Scholarly Texts," 100-106.

briefly consider evidence from Late Antiquity. In Late Antiquity, Aramaic was no longer only the vernacular of Babylonian culture but had also become the literary language of Mesopotamian Christianity, Babylonian Judaism and Mandaean Gnosticism.[66]

Considering the elements of Babylonian learning in Late Antique sources one may observe that these differ from what one finds in earlier cuneiform learning: Late Antique scholarship is different in character from the advanced mathematical astronomical texts from the latter part of the first millennium BCE, being less technical and more divinatory.[67] The textual similarities are of a more structural than of a specific nature (see section 4 below). What is important for our discussion is that these Late Antique traditions do not presuppose direct contact with cuneiform culture. The Late Antique texts do not derive from the "high culture" of cuneiform written tradition, but instead probably represent the transmission of elements of popular Babylonian traditions, although elite and popular culture should not be seen as two completely separate strands of tradition.[68]

[66] Cf. Beaulieu, "Official and Vernacular Languages"; Westenholz, "The Graeco-Babyloniaca Once Again," 307-308.
[67] This assessment may distract rather than explain. It is not meant as a normative observation but rather to indicate that Late Antique Babylonian sources do not transmit advanced mathematical astronomy but instead the divinatory astrological traditions. It is readily admitted that one goes with the other, as it did in first millennium BCE Babylonia and in Hellenistic astronomy and astrology, as in the case of Ptolemy. At the same time, when we find the divinatory side of the coin, we need not always presuppose knowledge of the technical and mathematical astronomical requirements. A telling example from the modern world comes from present-day India, which vividly shows how traditional horoscopy is practiced without mathematical astronomical knowledge; see Popović, *Reading the Human Body*, 162-163.
[68] See Westenholz, "The Graeco-Babyloniaca Once Again," 308. The situation of Late Antique Babylonian astrological literature is actually somewhat more complex, as some of the sources of these texts are not just Babylonian, but also Greek; see Popović, *Reading the Human Body*, 37, 111. Aage Westenholz's remark is directed at the medical, astrological and magical traditions reflected in the Talmud and the Mandaic texts from Late Antiquity, arguing that these do not derive "directly" from scholarly cuneiform tablets. The case with the texts from Qumran is different in that those texts are not examples of "popular" culture but of a different level of scholarly culture than that of cuneiform culture at the time. The point, however, is the same: neither derives directly from cuneiform sources. For the early Jewish evidence we likewise need to posit indirect channels of diffusion (see below).

There is, of course, some evidence for literary Aramaic texts from the first millennium BCE, and the Aramaic manuscripts from Qumran add significantly to this,[69] but this concerns literature (Aḫiqar)[70] or liturgy,[71] rather than Babylonian astronomy and astrology in its advanced forms. In our analysis of cultural encounters we need to distinguish between different kinds of texts and traditions and differentiate between various channels and agents of transmission.[72] What applies to one need not explain the other. For example, knowledge of the Gilgamesh epic concerns certain motifs that need not presuppose acquaintance with the Standard Version (not even consciously being related anymore to the epic as such),[73] but may have been part of popular oral traditions. Those who composed the Prayer of Nabonidus (4Q242) or Daniel 4 may have gotten their knowledge about the Babylonian king Nabonidus through a chain of transmission that originated in the public reading of Nabonidus' inscription.[74] In both cases, there is no need to suppose direct access to Babylonian centres of higher learning.

[69] See, e.g., Katell Berthelot and Daniel Stökl Ben Ezra (eds.), *Aramaica Qumranica* (STDJ 94; Leiden: Brill, 2010). See also Pollock, "Power and Culture Beyond Ideology and Identity," 285-286.
[70] See n. 58 above.
[71] See Richard C. Steiner, "The Aramaic Texts in Demotic Script: The Liturgy of a New Year's Festival Imported from Bethel to Syene by Exiles from Rash," *JAOS* 111 (1991): 362-363; idem, "Papyrus Amherst 63: A New Source for the Language, Literature, Religion, and History of the Arameans," in *Studia Aramaica: New Sources and New Approaches* (ed. M.J. Geller, J.C. Greenfield, and M.P. Weitzman; Oxford: Oxford University Press, 1995), 199–207.
[72] Cf. also Popović, "The Emergence of Aramaic and Hebrew Scholarly Texts," 109-110. On different cuneiform sources from ancient Israel, see Wayne Horowitz, Takayoshi Oshima, and Seth L. Sanders, *Cuneiform in Canaan: Cuneiform Source from the Land of Israel in Ancient Times* (Jerusalem: Israel Exploration Society, 2006).
[73] In addition to the references in n. 8 above, see Matthew Goff, "Gilgamesh the Giant: The Qumran *Book of Giants'* Appropriation of *Gilgamesh* Motifs," *DSD* 16 (2009): 221-253; André Lemaire, "Nabonide et Gilgamesh: L'araméen en Mésopotamie et à Qoumrân," in *Aramaica Qumranica*, 125-144; Ida Fröhlich, "Enmeduranki and Gilgamesh: Mesopotamian Figures in Aramaic Enoch Traditions," in *A Teacher for All Generations: Essays in Honor of James C. VanderKam* (Leiden: Brill, 2012), 637-653.
[74] Carol A. Newsom, "Why Nabonidus? Excavating Traditions from Qumran, the Hebrew Bible, and Neo-Babylonian Sources," in *The Dead Sea Scrolls* (ed. S. Metso, H. Najman, and E. Schuller), 57–79. See also Lemaire, "Nabonide et Gilgamesh."

Another kind of tradition is that of law and legal formulas in documentary texts, the origins of which have been traced back to Babylonia, such as in the Wadi Daliyeh papyri from fourth-century BCE Samaria and also in second-century CE legal texts from Murabbaʿat and Naḥal Ḥever.[75] The presence of empires as a political factor may account for the spread of legal traditions and formulary, and in this case the Aramaic *sepīru* scribes, taking care of all sorts of legal, administrative writings, may too have been agents conducive for the transmission of Babylonian legal formulas.

A differentiated perspective on the use of Aramaic in the ancient Near East points to different means of transmission, via different channels and agents, for different kinds of traditions and texts. What may apply to the transmission of elements from narratives, may not apply likewise to scholarly literature. That some astronomical elements in the *Astronomical Book* have Babylonian origins is evident, as do other elements in 1 Enoch.[76] This is not disputed. The issue is how these astronomical elements reached Jewish scholars and materialized in the extant manuscripts from Qumran. The suggestion put forward here is that in this respect these texts are analogous to Late Antique traditions in which Babylonian elements, and Greek ones, appear. For early Jewish astronomical and astrological traditions there is no

[75] See Douglas M. Gropp, "Wadi ed-Daliyeh: Written Material," in *Encyclopedia of the Dead Sea Scrolls* (ed. L.H. Schiffman and J.C. VanderKam; New York: Oxford University Press, 2000), 162-165 (163-164); D.M. Gropp et al., *Wadi Daliyeh II: The Samaria Papyri from Wadi Daliyeh and Qumran Cave 4.XXVIII: Miscellanea, Part 2* (DJD 28; Oxford: Clarendon, 2001), 5, 19–32. For these texts, see also recently Jan Dušek, *Les manuscrits araméens du Wadi Daliyeh et la Samarie vers 450-332 av. J.-C.* (Leiden: Brill, 2007); Jacobine C. Oudshoorn, *The Relationship between Roman and Local Law in the Babatha and Salome Komaise Archives: General Analysis and Three Case Studies on the Law of Succession, Guardianship and Marriage* (Leiden: Brill, 2007); Bernard S. Jackson and Daniela Piattelli, "A Recent Study of the Babatha and Salome Archives," *Review of Rabbinic Judaism* 13 (2010): 88–125.

[76] For example, familiarity in 1 Enoch with elements from the Enmeduranki tradition does not imply direct access to cuneiform sources. Such elements of cultural transmission are more diffuse, and therefore more difficult to pinpoint. This also applies to certain cosmographic and geographic elements in 1 Enoch with parallels in Babylonian and Greek traditions; see George W.E. Nickelsburg, *1 Enoch 1: A Commentary on the Book of 1 Enoch, Chapters 1-36; 81-108* (Hermeneia; Minneapolis: Fortress, 2001), 61–62, 279-289; Kelley Coblentz Bautch, *A Study of the Geography of 1 Enoch 17-19: "No One Has Seen What I Have Seen"* (Leiden: Brill, 2003), 231-257.

reason to suppose direct contact with Babylonian scholars or direct access to cuneiform texts, as we must assume for the Greek evidence.

3.3 The Transmission of Babylonian Astronomical Material to the Greek World

The exact routes by which Babylonian astronomical learning was transmitted to the Greek world are not known, but different suggestions have been put forward. We probably have to reckon with different channels of transmission that were in operation, sometimes perhaps simultaneously and sometimes not. What seems certain is that it must have involved, at one time at least, direct personal contact between individual Babylonian and Greek scholars, because of the technical and mathematical astronomical requirements, and even because of the visual aspects of the manuscripts such as their layout and structure. This may have happened when Greek astronomers visited Babylonia or when Babylonian scholars travelled to the Greek world. But we do not know exactly when, where and how this happened.[77] Furthermore, once this contact was established, a continuous tradition ensured its transmission until the days of Ptolemy and later.[78] Let us briefly consider these three channels of transmission: direct contact in Babylonia, direct contact elsewhere and a continuous tradition.

The first two proposed channels of transmission operate on the basis of the conclusion that direct access to cuneiform sources through contact between Greek and Babylonian astronomers must have occurred. For example, if, as has been suggested, the Greek astronomer Hipparchus in the second century BCE was the main channel through which important aspects of Babylonian astronomical science were transmitted to the Greeks, he must have had personal contact with one or more Babylonian experts. Hipparchus' extensive knowledge of both the observational and theoretical parts of Babylonian astronomy, it appears, cannot be explained without him having had direct access to Babylonian sources, most notably the observational Diaries.[79] If indeed Hipparchus personally accessed astronomical cuneiform sources with

[77] Jones, "The Adaptation of Babylonian Methods," 443; Steele, "Applied Historical Astronomy," 338-346; idem, "Visual Aspects," 454.
[78] Jones, "The Adaptation of Babylonian Methods," 443-444.
[79] See Toomer, "Hipparchus and Babylonian Astronomy," 357-360.

the help of one or more Babylonian scholars, we do not know whether he met them in Babylonia or elsewhere.

Although we do not know exactly the numbers journeying, and wherefrom and whereto they did so, Babylonian scholars certainly did travel and took up residence elsewhere in the Greek world, transmitting knowledge about Babylonian astronomy and astrology along the way. Scribes who were on the move and took their professional and scientific knowledge elsewhere would have been a potent link in the diffusion of knowledge.[80] For the Hellenistic period, the status that Babylonian astronomy and astrology enjoyed in the Greek world may have facilitated the reception of travelling scribes from Babylonia by local communities: knowledge of Babylonian astronomy crossed network distance and its effect on small-world networks was ensured by the status it already had. There is some intriguing documentary evidence for the diffusion of astronomical learning from Babylonia to the Greek world. An honorary inscription from second-century BCE Larissa in Thessaly refers to a certain Antipater from Hierapolis in Seleukia as a Chaldean astronomer who had lived for a long time in Larissa and had practised his mathematical profession there.[81] We do not know what kind of astronomy or astrology this Antipater pursued, but the inscription presents concrete evidence for a "Chaldean" astronomer/astrologer who had travelled from his native city in the Near East to a Greek city sometime in the second century BCE[82] If so, Antipater's "biography" illustrates that he crossed a large network distance and established himself successfully within his new social network, as testified by the honorary inscription.

[80] It has been argued for the Middle Babylonian period and the diffusion of scribal learning to Hittite Anatolia that the connections of travelling scribes to their target networks may have been superficially tenuous but that their effects on those target groups were often out of any proportion to the character of those connections. See Mark Weeden, "Adapting to New Contexts: Cuneiform in Anatolia," in *The Oxford Handbook of Cuneiform Culture*, 597–617 (603).

[81] SEG 31.576; C.J. Gallis, "New Inscriptions from Larissa," *Athens Annals of Archaeology* 13/2 (1980): 246-261 (250-251) (in Greek; English summary on pp. 261-262); Glenn W. Bowersock, "Antipater Chaldaeus," *CQ* 33 (1983): 491. I am grateful to Alexander Jones for the reference to this inscription and the two articles.

[82] The reference to Antipater having spent many years in Larissa (lines 11–12) may perhaps suggest that he came there as a young man.

How exactly he crossed this distance, through what channels and through what connecting localities, we do not know.[83] There is in any case no reason to assume that this Antipater from Hierapolis was the only astronomer to have moved away from his original social network. There were, no doubt, others like him who travelled to the Greek world (for the suggestion that astronomers from Uruk were among them, see section 3.1 above).[84]

In addition to these two channels of transmission of Babylonian astronomy to the Greek world via direct contact, a third one presupposes a continuous tradition once Babylonian science had been taken elsewhere. The use of Babylonian predictive schemes in the time of Ptolemy (second century CE) cannot be explained by direct contact with Babylonian scholars or direct access to cuneiform sources (Ptolemy received this information via Hipparchus).[85] A second-century CE fragmentary papyrus from Oxyrhynchus (P. Oxy 4139) has a reference to a lunar period scheme according to the Babylonian System A lunar theory and seems to refer to the Orchenoi ("people of Uruk"; line 8), indicating the diffusion of this Babylonian tradition.[86]

Considering the moment of "first contact," elements of Babylonian mathematical astronomy will have been transmitted to the Greek world directly from cuneiform into Greek, via personal contact with Babylonian astronomers, either in Babylonia or elsewhere. The Jewish world, on the other hand, as argued above, encountered some elements of Babylonian non-mathematical astronomy indirectly via the medium of the Aramaic language. We thus need to differentiate between channels for the westward transmission of Babylonian astronomy and astrology in their varied forms: directly from cuneiform into Greek and indirectly via Aramaic.

[83] Did he come via the Greek island of Cos? See Vitruvius, *On architecture* 9.6.2 and Bowersock, "Antipater Chaldaeus."

[84] Despite Berossus' reputation (founding a Chaldean school on the Greek island of Cos; Vitruvius, *On architecture* 9.6.2) scarcely anything connected with him is really astronomical. See Popović, *Reading the Human Body*, 217 n. 25. But see Amélie Kuhrt, "Berossus' *Babyloniaka* and Seleucid Rule in Babylonia," in *Hellenism in the East*, 32–56 (36–44).

[85] Jones, "The Adaptation of Babylonian Methods," 443-444.

[86] Jones, *Astronomical Papyri from Oxyrhynchus*, 1:97–99; 2:22–23.

4. The Transmission of Babylonian Astronomical and Astrological Learning to the Jewish World

In order to explain the Babylonian origins of some astronomical and astrological elements in early Jewish texts, we need not envisage direct access to cuneiform sources in Babylonian centres of learning. The Jewish evidence differs markedly from the Greek evidence, where such direct contact must be presumed. In addition to the different levels of learning displayed by the Babylonian, Greek and Jewish evidence, we should again consider the textual similarities between Jewish and Babylonian texts.

The Hebrew Qumran calendar texts (4Q320–321a) describe a consistent pattern of days between two phenomena from which it is clear that these phenomena are connected with the lunar cycle. This pattern has been explained by taking recourse to the Lunar Three scheme from Late Babylonian non-mathematical astronomical texts (see section 2.1 above), but this interpretation is not without difficulties.[87] Even if it would be certain that some elements of the Late Babylonian Lunar Three scheme were behind the Qumran calendar texts, these have been reworked in such a manner that there is no direct connection with actual Lunar Three texts.

Likewise, we need not suppose that the Aramaic *Astronomical Book* is a direct translation or rendering of cuneiform sources.[88] First, the texts do not look the same: Enūma Anu Enlil is shorter and more formulaic.[89] Second, similarities between the texts are of a structural rather than a specific nature.[90] As soon as a few astronomical and arithmetic principles from Enūma Anu Enlil Tablet 14 were understood

[87] Cf. Drawnel, *The Aramaic Astronomical Book (4Q208-4Q211) from Qumran*, 52 n. 170.

[88] See, however, Drawnel, *The Aramaic Astronomical Book (4Q208-4Q211) from Qumran*, 301, 304.

[89] See Farouq N.H. Al-Rawi and Andrew R. George, "Enūma Anu Enlil XIV and Other Early Astronomical Tables," *AfO* 38/39 (1991–1992): 52–69.

[90] Drawnel, *The Aramaic Astronomical Book (4Q208-4Q211) from Qumran*, 304-305 notes that both texts (1) present the numerical data in short formulaic sentences (although the cuneiform is much more terse than the Aramaic); (2) deal with visibility of the moon, presenting this in the scheme of one month; (3) compute the same time intervals of the night, although the Aramaic *Astronomical Book* inverts the order of the presentation in Enūma Anu Enlil; (4) divide night-time into two parts.

it would not have been difficult to transmit that knowledge further. One need not assume further direct access to the text of Enūma Anu Enlil itself. This also applies to the linear zigzag function that is rooted in the type of astronomy of Enūma Anu Enlil and Mul.Apin. The occurrence of such principles alone in texts from other periods and from other cultural realms need not assume direct access to the cuneiform sources in Babylonia. Once such learning was carried elsewhere by scribes trained in it, those principles may have become part of a continuous tradition through which other individuals gained access to it, either through contact with travelling scribes or further down the chain of transmission via other individuals who equally no longer had access to the cuneiform texts.

The astrological texts from Qumran (4Q186 and 4Q318) also do not presuppose direct access to cuneiform sources or direct contact with Babylonian scholars. Moreover, the zodiac names in 4Q318 and the reference to the horoscope in 4Q186 point to Hellenistic traditions (see section 2.1 above). In this respect, these texts are analogous to Late Antique traditions in which Babylonian and Greek elements appear (see section 3.2 above). Knowledge of the zodiac did not therefore reach Jewish Palestine directly via access to cuneiform sources. If we take the evidence for the Hellenistic origin of the zodiac names in 4Q318 into account, perhaps the chain of transmission went via Aramaic to Greek and back again, or, after the zodiac was already appropriated by the Greeks, via Greek to Aramaic, through which Jews in Palestine may have encountered it in the second or first century BCE and also translated it into Hebrew.[91]

The textual similarities between Jewish astronomical and astrological texts of the Hellenistic and early Roman periods and first millennium cuneiform sources from Babylonia are not of such a nature to suppose that the former are a direct translation or rendering of the latter. The similarities are 'structural' rather than specific. They concern general notions or motifs, rather than technical or specific

[91] On this 'translation,' see Popović, "The Emergence of Aramaic and Hebrew Scholarly Texts," 104.

data or visual layout and structure, as with evidence from the Greek world.[92]

Whether we should see this Jewish tradition as popular or elite depends on the perspective taken. In their own society these Jewish scholars may have been part of the elite, but given the enormous difference between this learning and the advanced mathematical astronomy that was exchanged between Babylonian and Greek scholars at that time, the Jewish scholars probably were not part of that same scholarly network.

Regarding the assumption that Jewish scholars had direct access to Babylonian learning, we cannot simply invoke the Babylonian "stream of tradition" as a fixed and stable entity available everywhere,[93] into which Jewish individuals could simply have tapped, almost at will, at certain moments in history, such as during the Babylonian exile or thereafter. Recent research on a number of cuneiform scholarly collections indicate that the "stream of tradition" was not simply present everywhere and available to everyone. The production, transmission and dissemination of scholarly knowledge and texts was conditioned by specific circumstances, thus affecting the concrete manifestations of the Babylonian "stream of tradition" at certain places and times.[94] In addition, prohibitions on tablet movement expressed through secrecy formulae, which appear as important topoi in the colophons of scholarly texts, may not have been absolute,[95] but some form of boundary maintenance with regard

[92] Cf. Westenholz, "The Graeco-Babyloniaca Once Again," 308, and see section 3.2 above.

[93] A. Leo Oppenheim, *Ancient Mesopotamia: Portrait of a Dead Civilization* (rev. ed. by E. Reiner; Chicago: The University of Chicago Press, 1977), 13, conceived of this "stream of tradition" as a stable corpus of scholarly writings that was relatively accessible to all learned men; it represented a cultural continuum that was maintained effectively by the scribal tradition.

[94] See Robson, "Empirical Scholarship"; eadem, "The Production and Dissemination of Scholarly Knowledge"; eadem, "Reading the Libraries"; eadem, "Tracing Networks of Cuneiform Scholarship." I am grateful to Eleanor Robson for providing me with copies of her articles before publication.

[95] It has long been debated in Assyriological studies how such secrecy and curse formulae should be understood. Recently, the pendulum seems to have swung back in the direction of the secrecy formulae being taken more or less at face value, implying that the scholarly corpora of astrologers, diviners and others were considered to be secret in principle. How this

to literary and scholarly texts did seem to have been in effect in Babylonia, limiting the accessibility and mobility of scientific knowledge.

Furthermore, the impression gained from cuneiform texts from the second half of the first millennium BCE in Babylonia is one of Babylonian urban elites who seemed reluctant to acknowledge the role of Aramaic and instead kept to traditional cuneiform as the sole official culture of Babylonia. The evidence is skewed because the cuneiform texts serve the self-interests of the Babylonian elite, conjuring up a view of Babylonia as an entirely cuneiform culture (see sections 3.1 and 3.2 above). However, from such an assessment we cannot infer that despite the Babylonian elite's self-interest and their preoccupation with their self-understanding amid a changing world there must have been a considerable body of Babylonian Aramaic literature and scholarship, all evidence of which subsequently was suppressed or must have vanished.

While the prestige of Babylonian astronomy facilitated the reception of Babylonian astronomers in the Greek world (see section 3.3 above), there may perhaps have been a similar effect in some cases with regard to Greek culture in the Babylonian world.[96] The Babylonian urban elite, with its thoroughly cuneiform culture and identity, seems to have positioned itself differently with regard to Greek culture and language than it did with regard to Aramaic culture and language. Some Babylonians learned Greek and participated in Greek scholarly networks.[97] This is not only illustrated by evidence for the direct transmission of advanced Babylonian mathematical astronomy to the Greek world, but also by Berossus and the so-called

principle will have worked out in specific contexts, however, is a different matter. The notion of secrecy and especially its social functions should be further specified in its socio-cultural contexts. See, e.g., Popović, *Reading the Human Body*, 81–83, 100, 227-231; Alan Lenzi, *Secrecy and the Gods: Secret Knowledge in Ancient Mesopotamia and Biblical Israel* (Helsinki: The Neo-Assyrian Text Corpus Project, 2008); Niek Veldhuis, "The Theory of Knowledge and the Practice of Celestial Divination," in *Divination and Interpretation of Signs in the Ancient World* (ed. A. Annus; Chicago: The Oriental Institute, 2010), 77–91 (79–80).

[96] See van der Spek, "Multi-Ethnicity and Ethnic Segregation," 111 for examples of non-peaceful coexistence.

[97] Cf. Beaulieu, "Official and Vernacular Languages," 210-211.

Graeco-Babyloniaca. The Graeco-Babyloniaca tablets (Akkadian written in Greek characters) from around the turn of the Common Era demonstrate that native Babylonians, students in this case, knew Greek.[98] Berossus represents a fascinating example of a native Babylonian who aimed to present his culture to the wider Hellenistic world beyond the community of traditional Babylonian learning and therefore wrote in Greek.[99] That he wrote in Greek and not in Aramaic seems significant. Of course, this was motivated by his wish to address a Greek-speaking audience, but this is precisely why it seems significant that he wrote in Greek. When some members of the Babylonian urban elite opened up, so to say, it was to Greek culture and language, something they had apparently not done in a similar manner earlier in the context of the Persian Empire with regard to the Aramaic language in relation to astronomy and astrology, Aramaic being only the vernacular.[100] Greek culture and language may have enhanced the self-esteem and identity of Babylonian elites, which Aramaic presumably did not in first-millennium BCE Babylonia.[101] In the Hellenistic period, therefore, the available sources seem to indicate that some of the Babylonian elite had an active interest in disseminating elements of their culture to non-Babylonians—but not just to any non-Babylonians.

While the evidence for the involvement of non-Babylonians in Babylonian science is meagre at best, the transmission of Babylonian astronomy to the Greek world and the cases of Berossus and the Graeco-Babyloniaca seem to suggest a different attitude among members of the Babylonian elite toward the Greeks. These

[98] For a recent discussion of these texts and references to other scholars and previous discussions, see Westenholz, "The Graeco-Babyloniaca Once Again."

[99] See most recently De Breucker, "Berossos between Tradition and Innovation."

[100] This changed attitude may perhaps have been caused by the extensive colonization programme started by the Hellenistic empires; see van der Spek, "Multi-Ethnicity and Ethnic Segregation," 106.

[101] Cf. Beaulieu, "Official and Vernacular Languages," 211. On the languages spoken by the Babylonian elites in the latter part of the first millennium BCE, see also, e.g., Geller, "The Last Wedge"; idem, "Graeco-Babyloniaca in Babylon," in *Babylon: Focus mesopotamischer Geschichte, Wiege früher Gelehrsamkeit, Mythos in der Moderne* (ed. J. Renger; Saarbrücken: SDV, 1999), 377-383; Westenholz, "The Graeco-Babyloniaca Once Again."

considerations argue against the assumption that Babylonian elites would have been actively involved in transmitting some of their astronomical knowledge to Jews who were, relatively, only quite recently in their midst,[102] as opposed to the Aramaeans who settled there earlier but against whom, it appears, it was possible to maintain a persistent sense of difference up until the Hellenistic period (even if we assume Jews to have been part of a general Aramaic milieu).[103] It seems unlikely that in Babylonia, where impressive developments took place in mathematical astronomy from the mid-first-millennium BCE onward, Babylonian scholars would have engaged in the transmission of older forms of their astronomy to local Jews or more recent insights if the Lunar Three scheme should indeed be seen as having influenced Qumran calendar texts.

Of course, we may posit all sorts of suppositions: that there must already have been a whole body of Babylonian literature and scholarship in Aramaic in the first millennium BCE; that Jews must have been active in administrative functions; that they knew cuneiform; and that they had access to Babylonian cuneiform literature and scholarship. But there is no evidence for this and also no reason to assume such conditions in order to explain the Jewish evidence for scholarly traditions. If we take the actual Babylonian elements into account, these amount to features that do not presuppose direct access to cuneiform sources. The connecting chains of early Jewish astronomical and astrological traditions were probably much more remote from the centres of Babylonian science than previously assumed. All this does not deny the Babylonian origins of elements in early Jewish astronomical and astrological traditions, but it does highlight that these were encountered via different channels and agents than the transmission of Babylonian astronomical science to the Greek world. The transmission of cuneiform culture in Aramaic in Late Antiquity may suggest one such different trajectory.

So how did Jewish scholars become acquainted with some of the learning originating with Babylonian scholars? What network connections and channels of transmission can we suggest for Jewish

[102] Cf. also Boiy, *Late Achaemenid and Hellenistic Babylon*, 187, 295.
[103] See n. 45 above.

scholars to have gained access to certain elements of Babylonian learning? Bearing in mind the three possible channels of transmission to the Greek world (see section 3.3 above), the one channel of transmission ruled out on the basis of the evidence available is that of direct contact in Babylonia between Jewish and Babylonian scholars. This leaves us with two possibilities, namely directly via contact with travelling scribes outside Babylonia or indirectly via a continuous tradition that had been transmitted to various localities through such travelling scholars. It is impossible to be specific as to the exact channels, localities and agents of transmission, as concrete evidence is lacking, but we might venture a hypothesis.

Two possible historical contexts seem to present themselves for obvious—but different—reasons: the Neo-Assyrian period and the Neo- and Late-Babylonian periods. Once we realize that for the "how" we need not assume direct contact, it is unnecessary to limit the cultural transfer of Mul.Apin and Enūma Anu Enlil type of astronomy to the Neo-Babylonian period or later.[104] Moreover, elements of Babylonian as well as Greek astronomical and astrological learning will have been transmitted at different moments in history, but there is no reason to think of these as clearly separate waves.[105] Logical necessity and historical sequence are not always identical in the history of science.[106] Thus, the older type of Enochic astronomy was used and slightly updated in Hebrew texts (4Q317, 4Q503 and possibly 4Q334) at a time when presumably other lunar schemes were also appropriated in the second century BCE (see section 2.1 above).

Only in the Neo-Assyrian period do we find evidence, at least in correspondence between scholars and the king, for the use of Aramaic in scholarly settings. The Neo-Assyrian period may thus have been conducive for the transmission of elements of cuneiform astronomical learning into Aramaic, from whence Aramaic became the medium for further diffusion of this material. This may have happened at the centres of learning, or perhaps at more provincial or peripheral localities. In addition, international scholars at the Neo-Assyrian court

[104] Ben-Dov, *Head of All Years*, 245-247.
[105] Ben-Dov, *Head of All Years*, 245-250.
[106] Walter Burkert, *Lore and Science in Ancient Pythagoreanism* (Cambridge, Mass.: Harvard University Press, 1972), 303.

may have taken such traditions with them on their travels and transmitted elements of them into Aramaic along the way. Via such an indirect trajectory and through various points in between it may have reached Jews in Palestine sometime in the Hellenistic period, or perhaps already in the Persian period. As with the Late Antique evidence, we need not assume direct contact to explain the early Jewish evidence and can thus allow for a certain amount of time for such a hypothesized continuous tradition and for some of its elements to have materialized in the Aramaic *Astronomical Book* and Enochic astronomy.

If we posit the Neo- or Late-Babylonian periods as the historical context for this diffusion, the Aramaic *Astronomical Book* from Qumran is evidence for the transmission of Babylonian learning via the medium of Aramaic in first-millennium BCE Babylonia. This transmission probably did not happen in a direct manner such that Jews learned this type of astronomy through personal interaction with Babylonian or Assyrian scholars in their centres of learning. There being no reason to see the origin of the Aramaic *Astronomical Book* in the eastern Diaspora, it should be considered a Jewish Palestinian text. It is, therefore, not necessary to posit Babylonia as the place where the actual transmission to Jewish individuals must have occurred.[107]

The astrological texts from Qumran seem to point to a different channel of transmission because of the Hellenistic astrological elements alongside a Babylonian one in the case of 4Q318 and probably 4Q186. These texts do not presuppose direct contact with cuneiform sources or Babylonian scholars, but may point to a more vague, continuous tradition of astronomical and astrological lore, analogous to Late Antique traditions (see above in this section). As for the Qumran calendar texts, if indeed these did have elements of the Late-Babylonian Lunar Three scheme as their model of inspiration, it seems significant that we find them rendered directly in Hebrew

[107] See, however, Drawnel, *The Aramaic Astronomical Book (4Q208-4Q211) from Qumran*, 53.

rather than Aramaic.[108] This argues against direct access to cuneiform sources and contact with Babylonian scholars.

5. Conclusion

Contrary to what is often assumed,[109] there is no evidence to suggest that during the first millennium BCE Jewish scholars had direct access to Babylonian centres of higher learning, such as the temples in Neo- and Late-Babylonian times, where they interacted with Babylonian astronomers and were able to read with their help cuneiform astronomical texts and appropriate such learning.

We should no longer think of a disembodied Babylonian "stream of tradition" that was simply out there and available and which individual representatives of Jewish tradition could have accessed effortlessly. Considering cultural encounters between Jews and Babylonians, there would have been different aspects involved and different levels of interaction to reckon with. Explanations for the appropriation of motifs from Gilgamesh, for acquaintance with Nabonidus traditions, for the reception of legal formulas or for the transmission of astronomical and astrological traditions should not be lumped together indiscriminately as being the result of direct access to elite Babylonian learning. In our approach we need to distinguish between different kinds of texts and traditions and differentiate between various channels and agents of transmission.

Regarding the transmission of Babylonian astronomical and astrological material to the Jewish world, we have considered the transmission of advanced Babylonian mathematical astronomy into Greek sources, which suggests high-level contact between Greek and Babylonian scholars, either in Babylonia or elsewhere, or, at a later date when direct contact would not have been possible anymore, through a continuous tradition. Jewish scholars, at least as far as our sources are concerned (setting aside indirect testimonies, such as Abraham being an astronomical teacher, or the Daniel narratives

[108] Or should we assume Aramaic intermediaries, as with 4Q186 and 4Q561 (although not in a strict sense)? Cf. Popović, "The Emergence of Aramaic and Hebrew Scholarly Texts," 103–6. Jonathan Ben-Dov, "Scientific Writings in Aramaic and Hebrew at Qumran: Translation and Concealment," in *Aramaica Qumranica*, 379–402 does not seem to reflect on this aspect.

[109] See the references in n. 2 above.

presenting a Jewish sage at the Babylonian court), were not part of that same high-level network: the Jewish sources testify to a different level of scholarship. Future research should therefore look for network connections at other places than the centres of Babylonian learning, as has often been the case. Nor should we perceive of the transmission of Babylonian astronomical and astrological material to Second Temple period Palestine as having taken place in a direct manner but rather through various intermediaries, Aramaic and other channels, as well as via a—more vague—continuous tradition.

Via such indirect channels, elements of Babylonian and Greek astronomical and astrological learning reached certain people and certain places in Jewish Palestine, at least those at Qumran and the movement behind the Dead Sea Scrolls. Regarding these people, one might suggest that they formed a rather closed network, which was not particularly interested in current developments in the outside world. However, the scientific material shows otherwise, indicating a degree of openness to scholarly learning from other traditions. The different level of learning in comparison with contemporary Babylonian and Hellenistic astronomy suggests that although "Qumran" was connected, it was not as well connected to those scholarly networks that participated in the transmission of knowledge between Babylonia and Greek culture in Greece, Egypt and elsewhere in the Hellenistic world.

In the centuries after the turn of the era things may have been different in terms of "high-level" contact between Jewish and Babylonian elites,[110] but these data cannot be anachronistically projected backward into the first millennium BCE Jewish participation in society may simply have been different in those changed historical

[110] See, e.g., Shai Secunda, "Talmudic Text and Iranian Context: On the Development of Two Talmudic Narratives," *AJS Review* 33 (2009): 45–69; idem, "Reading the Bavli in Iran," *JQR* 100 (2010): 310-342.

contexts.[111] The times of Medieval and especially early Modern Europe were still far away.[112]

[111] The difference in evidence, both quantitatively and qualitatively, for Jewish magic and astrology between the Second Temple period and Late Antiquity and the early Medieval period may perhaps also be indicative for such different involvement; see Popović, "Astrologische und magische Traditionen."

[112] See David B. Ruderman, *Jewish Thought and Scientific Discovery in Early Modern Europe* (New Haven: Yale University Press, 1995). For their suggestions and comments I thank the editors, the two anonymous reviewers, Jan Bremmer and Florentino García Martínez.

8. "Ancient Jewish Sciences" and the Historiography of Judaism

Annette Yoshiko Reed*

Even a decade or so ago, it might have been difficult to imagine an entire conference and volume devoted to "ancient Jewish sciences." In *Jewish Thought and Scientific Discovery in Early Modern Europe*, for instance, David Ruderman noted how he had "originally intended to begin ... with an overview of attitudes towards the natural world in ancient Judaism," only to encounter "a vast body of material in an area that has not been fully studied."[1] Such was the dearth of research that it warranted its own Appendix—a bibliographical essay sketching possible paths ahead.[2] Similarly, in 2002, when Y. Tzvi Langermann investigated the "beginnings of Hebrew scientific literature," the period that he had in mind was the eighth and ninth centuries CE, and the works in question were *Baraita de-Shmuel*, *Sefer Yetzirah*, *Mishnat ha-Midot*, and *Yetzirat ha-Walad*, as well as possible "products of the same creative spurt" such as *Pirqe de Rabbi Eliezer*, *Midrash Konen*, and *Sefer Asaf ha-Rofe*.[3] Shortly after, Shlomo Sela cited many of the same works as "early" precedents for the twelfth-century "rise of medieval Hebrew science," albeit stressing that even these can be "hardly

* Earlier versions of this essay were pre-circulated for and presented at the ISAW conference on which the present volume is based. I am grateful to Jonathan Ben-Dov, Seth Sanders, and Mladen Popović for thought-provoking conversations before, during, and after the conference; I hope that the present essay captures even a little of what I have learned from them. Benjamin Wright, Lawrence Schiffman, Seth Schwartz, Steven R. Reed, and Benjamin J. Fleming also offered crucial feedback on various earlier forms. Special thanks to David Ruderman for theoretical insights and bibliographical suggestions on the penultimate draft, and to William McCants and Nicholas Harris for their aid in navigating the relevant Islamic and Judeo-Arabic materials.

1 David B. Ruderman, *Jewish Thought and Scientific Discovery in Early Modern Europe* (New Haven: Yale University Press, 2001 [1995]), 375.

2 Ibid, 375-382.

3 Y. Tzvi Langermann "On the Beginnings of Hebrew Scientific Literature and on Studying History through 'Maqbilot' (Parallels)", *Aleph* 2 (2002): 169-176.

described as a homogenous corpus ... belonging to a continuous scientific tradition."[4]

The possibility of even earlier precedents was among the points that I sought to raise a few years later, in an essay bringing Second Temple and other Jewish sources to bear on broader historiographical problems pertaining to "religion" and "science."[5] Although materials of possible relevance had been discussed since the 1970s in specialist research on sapiential and apocalyptic traditions in Second Temple Judaism,[6] little had been done to bring them into the purview of the scholarly project of Ruderman and others who sought to forge "a new area of inquiry in Jewish cultural history" by "open[ing] a meaningful conversation about the dialogues between science and Judaism" in their shifting social, discursive, and disciplinary contexts.[7] And thus, even in 2007, it was still pertinent to ask "Was there science in ancient Judaism?"[8] and to begin an inquiry into the question by stressing that "[t]he topic of science in ancient Judaism has attracted surprisingly little scholarly attention."[9]

The difference, since then, is stunning. The possibility that evidence from the Dead Sea Scrolls might transform our views about this topic, as with so many others, was raised already in the wake of J. T. Milik's publication of the Aramaic fragments of 1 Enoch from

4 Shlomo Sela, *Abraham Ibn Ezra and the Rise of Medieval Hebrew Science* (Leiden: Brill, 2003), 2–3, 7.

5 Annette Yoshiko Reed, "Was there Science in Ancient Judaism? Historical and Cross-Cultural Reflections on 'Religion' and 'Science,'" *SR* 36 (2007): 461-495.

6 E.g., Michael E. Stone, "Lists of Revealed Things in the Apocalyptic Literature," in *Magnalia Dei: The Mighty Acts of God: Essays on the Bible and Archaeology in Memory of G. Ernest Wright* (ed. F. M. Cross, W. Lemke, and P. D. Miller; Garden City, N.Y.: Doubleday 1976), 414-452; Martha Himmelfarb, *Ascent to Heaven in Jewish and Christian Apocalypses* (New York: Oxford University Press, 1993), 72-94; Philip S. Alexander, "Enoch and the Beginnings of Jewish Interest in Natural Science," in *The Wisdom Texts from Qumran and the Development of Sapiential Thought* (reprinted in this volume, ch. 2).

7 Ruderman, *Jewish Thought*, xv. To be sure, Alexander, "Enoch and the Beginnings," cites Ruderman's work as inspiration, but its concern is less with the cultural or social history of knowledge, and more with cognitive shifts; see further below.

8 Jacob Neusner, "Why No Science in Judaism?," *Shofar* 6.3 (1988): 45–71.

9 Reed, "Was there Science in Ancient Judaism?," 461.

Qumran.[10] Yet, here as elsewhere, the full realization of the ramifications awaited the publication of the entire corpus, and the new syntheses thus made possible. In this sense, the very notion of "ancient Jewish sciences" owes much to two monumental monographs on the evidence from Qumran: Mladen Popović's *Reading the Human Body* in 2007 and Jonathan Ben-Dov's *Head of All Years* in 2008. By considering the full range of relevant Hebrew and Aramaic fragments discovered among the Dead Sea Scrolls in their broader Jewish, Mesopotamian, and Hellenistic contexts, Popović and Ben-Dov have opened startling new vistas onto the engagement of ancient Jewish scribes and scholars with the scientific discourses, discoveries, and disciplines of their non-Jewish counterparts.

Through a focus on 4Q186 and 4Q561, Popović[11] recovers a physiognomic tradition "the existence of which was unknown before the Qumran discovery": not only do these fragments reveal the surprising ways in which "Second Temple period Judaism participated in forms of learning that were current in surrounding cultures,"[12] but 4Q186 offers "exceptional textual evidence for physiognomic and astrological learning in antiquity in general" inasmuch as physiognomics is here used "to determine a person's zodiacal sign."[13]

[10] Józef T. Milik, *The Books of Enoch: Aramaic Fragments of Qumran Cave 4* (Oxford: Clarendon, 1976); Michael E. Stone, "Lists of Revealed Things in the Apocalyptic Literature"; James C. VanderKam, *Enoch and the Growth of an Apocalyptic Tradition* (CBQMS 16; Washington D.C.: Catholic Biblical Association of America, 1984), 76-109; Otto Neugebauer, "The 'Astronomical Chapters' of the Ethiopic Book of Enoch (72 to 82)", in Matthew Black, *The Book of Enoch or 1 Enoch: A New English Edition* (SVTP 7; Leiden: Brill, 1985), 386-414.

[11] Mladen Popović, "Physiognomic Knowledge in Qumran and Babylonia: Form, Interdisciplinarity, and Secrecy," *DSD* 13 (2006): 150-176; idem, *Reading the Human Body: Physiognomics and Astrology in the Dead Sea Scrolls and Hellenistic-Early Roman Period Judaism* (STDJ 67; Leiden: Brill, 2007).

[12] Popović, "Physiognomic Knowledge," 150-151; idem, *Reading the Human Body*, 68–118; cf. Markham Geller, "New Documents from the Dead Sea: Babylonian Science in Aramaic" in *Boundaries of the Ancient Near Eastern World: A Tribute to Cyrus H. Gordon* (ed. Meir Lubetski, Claire Gottlieb, and Sharon Keller; JSOTSup 273; Sheffield: Academic Press, 1998), 224-229.

[13] Popović, "Physiognomic Knowledge," 150-151, 165; idem, *Reading the Human Body*, 18–54, 119–208; cf. Philip S. Alexander, "Physiognomy, Initiation, and Rank in the Qumran Community," in *Geschichte—Tradition—Reflexion: Festschrift fur Martin Hengel zum 70. Geburtstag* (ed. H. Cancik, H. Lichtenberger, and P. Schäfer; Tübingen: Mohr Siebeck, 1996), 385-394;

Ben-Dov's comprehensive and comparative treatment of calendrical materials from Qumran achieves something similar with respect to astronomy.[14] Further establishing the value of the Dead Sea Scrolls as evidence for the intercultural connectivity of knowledge in the Hellenistic Near East, he stresses that ancient Jewish engagement with Babylonian astronomy was far more involved and sustained than previously assumed (e.g., with adoption of Mul.Apin-type astronomy, as well as concepts which arose subsequent to it).[15] At the same time, Ben-Dov makes a persuasive case for understanding Jewish appropriation of such knowledge as more than mere "borrowing," and he thus opens the way for a fresh approach to "ancient Jewish sciences" as a "unique amalgam" that can be analyzed "from within" as a "self-contained intellectual construct." [16]

In what follows, I would like to revisit some concerns from my earlier essay in the light of this new research. Here as there, my focus shall be less on the place of Judaism within the history of science, and more on the place of "science" in the history of Judaism. Accordingly, I shall not be concerned to argue for this or that Jewish text as "*really* scientific," whether by the standards of non-Jewish cultures of the past, or by the standards of modern ideals of rationalism, empiricism, secularism, or progress. My interest, rather, is in trying to recover some of the ways in which knowledge about the stars, cosmos, and human body was represented, taught, and transmitted in premodern Jewish literary cultures—particularly in relation to pedagogies and practices more distinctively marked as "Jewish" (e.g., halakha, biblical exegesis, parabiblical literary production).

What I shall suggest, in the first section of this essay, is that the case of ancient Judaism might provide an apt test-case for exploring the evolving place of cultural specificity in the perception and practice

Francis Schmidt, "Ancient Jewish Astrology: An Attempt to Interpret 4QCryptic (4Q186)", in *Biblical Perspectives: Early Use and Interpretation of the Bible in Light of the Dead Sea Scrolls* (ed. Michael E. Stone and Esther G. Chazon; STDJ 28; Leiden: Brill, 1998), 189-205.

[14] Jonathan Ben-Dov, *Head of All Years: Astronomy and Calendars at Qumran in their Ancient Context* (STDJ 78; Leiden: Brill, 2008), 4–5.

[15] Cf. Matthias Albani, *Astronomie und Schöpfungsglaube: Untersuchungen zum astronomischen Henochbuch* (WMANT 68; Neukirchen-Vluyn: Neukirchener Verlag, 1994).

[16] Ben-Dov, *Head of All Years*, 1, 276-278, 286-287.

of "science." I begin by considering Jewish attempts to insert Jews into Greek, Roman, and Islamic histories and taxonomies of knowledge. I point to continuities in the ancient and medieval Jewish appeal to biblical figures for such aims, but also to the character and degree of cross-cultural translatability thereby presumed. Jewish examples prove especially useful—I propose—for tracing the movement of knowledge across local and linguistic boundaries, and for illumining the shifting imperial contexts in which such knowledge found expression in written forms.

Yet attention to "science" might prove useful for the historiography of Judaism as well. In the second section, I argue that recent insights from the history of science help to highlight anachronistic assumptions about "religion" and "science" in research on ancient Judaism, while also modeling some alternate frames of analysis. Attention to "ancient Jewish sciences," in turn, may help us to navigate more integrative approaches to Jewish cultural history, wherein our data for ancient Judaism is not artificially reduced to modern notions of "religion."

Finally, in the third section, I sketch out some of the ramifications, particularly with respect to our understanding of diversity and continuity within Judaism. Not only does new work on "ancient Jewish sciences" offer new vantage-points on relevant Rabbinic materials, but it pushes us to reconsider possible continuities in Jewish reflection on the cosmos into Late Antiquity and the Middle Ages. Is the evidence of Hebrew fragments like 4Q186 or 4Q317, for instance, merely a curious footnote to Langermann's location of the "first chapter in the history of Hebrew scientific literature ... sometime during the very late Umayyad and early Abbassid caliphates," as an inner-Jewish reflex of the "surge of interest in science" in the Islamic imperial cultures of the time?[17] Might any lines of continuity connect the astronomical and cosmological concerns in the Enoch literature and at Qumran, with those that find expression at the end of Late Antiquity, in works like *Baraita de-Shmuel*, *Pirke de Rabbi Eliezer*, and *Seder Rabbah di-Bereshit*? Might some of these lines run through the classical Rabbinic literature as well? Could new insights into the most

[17] Langermann, "On the Beginnings," 170, 175.

ancient of known "ancient Jewish sciences" make it newly possible to imagine a history of Jewish engagement in sciences, as something more than a collection of disconnected instances of the engagement of individual Jews in the knowledge-enterprises of Babylonian, Hellenistic, Islamic, European, and other peoples?

1. *Cultural specificity and premodern heurematography*[18]

First, and by means of preface, it is perhaps important to ask: What do we even mean when we speak of "Jewish science,"[19] "Islamic science," "Babylonian science," or so forth? Isn't part of the point of speaking of "science" to signal types of knowledge that are not bound to the cultural specificities of their points of origin—those bodies of wisdom, ways of knowing, and aligned pedagogies, techniques, and technologies that remain falsifiable or utile, irrespective of creed and culture? Or, in other words: if we have to call it "ancient Jewish sciences," are we actually talking about "science" at all?

At first sight, it might seem easy enough to side-step such questions. Indeed, even today, "science" is far from a monolith, bearing coherence and unity mainly as an ideology shared by multiple distinct disciplines.[20] Research on its early modern history, moreover, has exposed the cultural contingency and constructedness of the very

[18] By "heurematography," here and below, I mean the practice of pinpointing and listing the discoverers or inventors of specific skills and knowledge; see K. Thraede, "Erfinder II," in *RAC* V: 1191-278; Leonid Zhmud, *The Origin of the History of Science in Classical Antiquity* (trans. Alexander Chernoglazov; Berlin: de Gruyter, 2006); William F. McCants, *Founding Gods, Inventing Nations: Conquest and Culture Myths from Antiquity to Islam* (Princeton: Princeton University Press, 2011).

[19] Inasmuch as my focus here is on premodern materials, I do not tackle the problem of the modern reception of disciplines like psychoanalysis as "Jewish sciences"—an issue that raises its own set of questions about "science" and cultural specificity, on which see further, e.g., Stephen Frosh, "Freud and Jewish Identity," *Theory & Psychology* 18.2 (2008): 167-178.

[20] Peter Dear, "What Is the History of Science the History Of? Early Modern Roots of the Ideology of Modern Science," *Isis* 96 (2005): 401-4.

category, as a product of specific intellectual and institutional developments in Europe, particularly in the nineteenth century.[21]

Yet the problem of the cultural specificity of knowledge still remains. After all, our modern Western concept of "science" may be culturally and historically specific, but—as scholars of colonial sciences have recently stressed[22]—its prehistory, production, and practice are inextricably embedded within a cross-cultural matrix. The rhetoric of universalism may mask the Eurocentrism of some modern ideologies of "science," but it also signals a notable self-consciousness concerning the consolidation and theorization of knowledge culled from an interreligious *oikoumenê*, fostered within transnational contexts, and promoted and naturalized within global networks to this day.[23]

Significantly, for our purposes, a similar dynamic seems to have been present in premodern times, and perhaps even acknowledged as such. Curiosity about the culturally-specific origins of various cross-culturally-diffused *technai* was among the hallmarks of ancient Greek ethnography, as first attested in the sixth century BCE, and consolidated as part of the reflection on "Greekness" and cultural difference during and after the Persian Wars.[24] Likewise, the Greek practice of heurematography functioned, in part, to telegraph

[21] Stanley Tambiah, *Magic, Science, Religion, and the Scope of Rationality* (Cambridge: Cambridge University Press, 1990), 8-18; John H. Brooke, *Science and Religion: Some Historical Perspectives* (Cambridge: Cambridge University Press, 1991), 52–81, 286-289; John Henry, *The Scientific Revolution and the Origins of Modern Science* (New York: Palgrave, 2002, 2nd ed.), 1–13.

[22] Roy MacLeod, "Introduction: Nature and Empire: Science and the Colonial Enterprise," *Osiris* 15 (2000): 1-13; David Wade Chambers and Richard Gillespie "Locality in the History of Science: Colonial Science, Technoscience, and Indigenous Knowledge," *Osiris* 15 (2000): 221-240.

[23] Marwa Elshakry suggests, in fact, that even the notion of "Western science" owes much to the reception of European ideas and disciplines among intellectuals in the Middle East and Asia; see Elshakry, "When Science Became Western: Historiographical Reflections," *Isis* 101.1 (2010): 98-109; also Lissa Roberts, "Situating Science in Global History: Local Exchanges and Networks of Circulation," *Itinerario* 33 (2009): 9-30. These recent approaches contrast with older models for the spread of scientific knowledge across cultures. Once dominant but now rejected, for instance, is the diffusionist model of George Basalla, which posited a three-stage process whereby pre-scientific, non-Western nations first provided resources for Western science, then embraced them, and finally used them as a basis for developing their own, national sciences; see, e.g., Basalla, "The Spread of Western Science," *Science* 156 (1967): 616-622.

world-histories, handily resolving the tensions between local and universal knowledge by mapping cultural difference along the axis of time: just as the ancient Greek category of *technai* encompassed the scientific, technical, and other skills deemed cross-culturally constituent of human civilization, so the listing of a single inventor or source-culture for each element thereof served to posit a common human past.[25]

That such commonality was constructed for a Greek gaze and present, however, made such discourses readily adaptable for later imperial contexts.[26] Indeed, it is perhaps not coincidental that the historiography of sciences first flourished in the immediate wake of the conquests of Alexander the Great. The Hellenistic empires of Alexander's successors seem to have facilitated the diffusion and development of scientific knowledge across an increasingly interconnected Mediterranean and Near East, but also a cross-cultural discourse about the origins and diffusion of such knowledge. Surviving fragments from Eudemus of Rhodes (ca. 330–285 BCE), for instance, attest an interest among Greeks in tracing the development of

[24] Adolf Kleingünter, *Protos Heuretes: Untersuchungen zur Geschichte einer Fragestellung* (Leipzig: Akademie-Verlag, 1933); Arther O. Lovejoy and George Boas, *Primitivism and Related Ideas in Antiquity* (Baltimore: John Hopkins University Press, 1997 [1935]), 382-388.

[25] Thraede, "Erfinder II," in *RAC* V; Zhmud, *Origin of the History of Science*, 10–54; Marcella Farioli, "The Genesis of the Cosmos, the Search for *Arche* and the Finding of *Aitia* in Classical Greek Culture," in *Origins as a Paradigm in the Sciences and Humanities* (ed. Paola Spinozzi and Alessandro Zironi; Göttingen: V&R Unipress, 2010), 195-209.

[26] I.e., with the importance of the local or culturally-distinct origins of certain disciplines predicated on (and thus subordinated to) their delocalized and trans-cultural value. This delocalized value, notably, is determined from an imperial vantage-point, the agency of which is erased by the assertion of a "universal" horizon for their diffusion and reception. In other words: what can appear to be a nativizing discourse about competing claims to antiquity and cultural priority of scientific discovery is often also (if more invisibly) participation in a totalizing discourse of empires. On the place of imperial power and knowledge in the spread, organization, and theorization of knowledge to and about Jews in the Hellenistic and Roman periods—as well as the differences in Hellenistic, Roman, and Islamic imperial strategies and stances—see Annette Yoshiko Reed, "Abraham as Chaldean Scientist and Father of the Jews: Josephus, *Ant.* 1.154–168, and the Greco-Roman Discourse about Astronomy/Astrology," *JSJ* 35 (2004): 145-156; McCants, *Founding Gods*.

mathematics and astronomy[27] already by the late fourth century BCE. Fragments from the writings of the Babylonian priest Berossus (ca. 330–280 BCE) and the Egyptian priest Manetho (fl. ca. 280–260 BCE) speak to a parallel development in the Hellenistic Near East shortly thereafter, whereby native elites wrote in Greek to mount competing claims to priority and antiquity in disciplines such as astronomy/astrology.[28]

The results of Greek ethnography, Hellenistic sciences, and Near Eastern competitive historiography were later appropriated by Roman elites as well. Even if pragmatic and political aims displaced the characteristically Greek concern with intellectual prestige,[29] the organization of knowledge from and about different peoples seems to have remained a powerful tool for articulating totalizing claims of empire. Within the Roman Empire and its Christian heirs, cultural difference continued to be conceptually managed, in part, through literary practices like scientific encyclopedism and universal history—with the promise of empire, to unite a multiplicity of locales in harmonious singularity, mirrored by the claim to comprehensiveness conveyed by the anthological forms of many of the literary genres used for explaining, organizing, or transmitting scientific knowledge.[30]

So too in later Islamic empires. The resurgence of interest in Hellenistic sciences in the Abbasid age, for instance, was accompanied by a literature on "firsts" [*awā'il*] that encompassed a variety of peoples and pasts, thus helping to mediate the imperial paradox of unity in diversity, while buttressing new claims to authority as the natural center and arbiter of human knowledge.[31] As in Hellenistic, Roman, Byzantine, and other empires before them (and modern European nation-states after them), the historiography of sciences

[27] Zhmud, *Origin of the History of Science*, 117-165.

[28] Arnaldo Momigliano, *Alien Wisdom: The Limits of Hellenization* (Cambridge: Cambridge University Press, 1975); Gregory Sterling, *Historiography and Self-Definition: Josephus, Luke-Acts, and Apologetic Historiography* (NovTSup 54; Leiden: Brill, 1992), 20-225.

[29] Momigliano, *Alien Wisdom*.

[30] Reed, "Abraham as Chaldean Scientist."

[31] Dimitri Gutas, *Greek Thought, Arabic Culture* (London: Routledge, 1998); John Walbridge, "Explaining Away the Greek Gods in Islam," *Journal of the History of Ideas* 59.3 (1998): 389–403; McCants, *Founding Gods*.

seems to have proved particularly powerful for promoting and naturalizing Abbasid claims to the status as true heir to the totality of cross-culturally valuable information discovered in the past and throughout the world.

Partly as a result, the problem of the cultural specificity of "science" has formed part of the theorization of the history of knowledge within Judaism as well. In ancient and medieval Jewish sources written under Hellenistic, Roman, and Islamic rule—in Greek and Arabic, as well as Hebrew and Aramaic—one finds assertions of participation or priority in the cross-cultural study of the stars, cosmos, and human body.[32] Such claims are sometimes articulated within the frameworks of the imperial discourses of harmonized difference that we find attested also in non-Jewish sources, and at other times, in terms that resist the totalizing rhetorics and epistemologies of ruling powers. What proves consistent, however, is the appeal to biblical figures, categories, and models. Writing in variety of genres, languages, and contexts, ancient and medieval Jews seem to have situated themselves within imperial and other cross-cultural histories of human knowledge primarily with reference to the narratives, heroes, and categories of the Torah/Pentateuch and other biblical traditions.[33]

Already at the dawn of the Hellenistic age, the Jewish scribes responsible for the Aramaic *Astronomical Book* and *Book of Watchers* (ca. third century BCE) point to the antediluvian sage Enoch (cf. Gen 5:18–24) as the ultimate source and conduit for true knowledge about the structure and workings of the cosmos.[34] In these most ancient

[32] Abraham Melamed, *The Myth of the Jewish Origins of Science and Philosophy* [Hebrew] (Jerusalem: Magnes, 2010).

[33] See, e.g., Reed, "Abraham as Chaldean Scientist," on Josephus, his Hellenistic Jewish predecessors, and his Roman contemporaries; Giuseppe Veltri, "The Rabbis and Pliny the Elder: Jewish and Greco-Roman Attitudes toward Magic and Empirical Knowledge," *Poetics Today* 19.1 (1998): 63-89, on late antique Rabbis and Pliny; Y. Tzvi Langermann, "Science and the Kuzari," *Science in Context* 10 (1997): 495–522, on Judah Halevi in his Islamic philosophical and scientific cultural contexts. For a comprehensive survey of the premodern articulations of the trope of Jewish priority in sciences in relation to parallel premodern claims about philosophy, as well as modern reticence about them, see now Melamed, *Myth of the Jewish Origins of Science*.

[34] Esp. 1 Enoch 17–36, 72–82; VanderKam, *Enoch*, 76–109; Himmelfarb, *Ascent to Heaven*, 72–94.

Enochic works, it is unclear whether Enoch is meant more to connect the histories of Jewish knowledge with those of other peoples, or more to distinguish them. Enoch's association with Babylonian astronomy in the *Astronomical Book*, for instance, may be articulated in deliberate resistance to the newer Hellenistic astronomy spreading at the time;[35] if so, it may signal a concern to assert true knowledge about the stars as preserved only or especially among the Jews. In the association of Enoch and his revelations with the hybrid Hellenistic-Babylonian geography and cosmology of the *Book of Watchers*,[36] by contrast, we may glimpse more of a concern to assert the ancient Jewish origins for ideas and disciplines well-known to be widespread across different cultures, thus establishing the place of the Jews in world-history and the debt of other cultures to their discoveries (or, rather—according to these works—to their mediation of heavenly knowledge down to earth). And perhaps, in both cases, the choice of a figure of such extreme antiquity as a spokesman for "ancient Jewish sciences" might signal some ambivalence about precisely these issues: Enoch, after all, lived prior to what the Torah/Pentateuch presents as Israel's genealogical distinction with Shem, its chosenness with Abraham, and its peoplehood with Moses, and long before what Greek historians even deemed recorded history.

Whatever else can be inferred from Enoch's presentation as *prôtos heuretês*, however, it remains that the earliest Enochic works exhibit an interest in tracing the genealogy of knowledge shared by contemporary authors like Berossus and Eudemus. Rather than lauding Babylonian archives or Greek sages as privileged loci for the history of human knowledge, the Jewish authors/redactors of the *Astronomical Book* and *Book of Watchers* promoted the literary heritage of Israel. Whether or not they sought primarily to stake a Jewish claim over the origins or character of sciences, their appeal to Enoch thus

35 Michael E. Stone, "Enoch, Aramaic Levi and Sectarian Origins," *JSJ* 19 (1988): 159-170.

36 Cf. Kelley Coblentz Bautch, *A Study of the Geography of 1 Enoch 17-19: "No One Has Seen What I Have Seen"* (JSJSup 81; Leiden: Brill, 2003); Annette Yoshiko Reed, "Enoch, Eden, and the Beginnings of Jewish Cosmography," in *The Cosmography of Paradise* (ed. Alessandro Scafi; London-Turin: The Warburg Institute-Nino Aragno, forthcoming).

functions to assert the unique place of Jewish texts and traditions for transmitting the earliest history of human knowledge.[37]

Shortly afterwards, in the second century BCE, the Book of Jubilees more decisively marks Enoch and his astronomical knowledge as "Jewish," by associating them with a line of books and teachings preserved solely by the descendants of Abraham, Jacob, and Levi, and articulated in contrast to the transmission of divinatory texts and teachings in lineage from the fallen angels and in languages other than Hebrew.[38] Yet Enoch is also conscripted for more cosmopolitan approaches to the place of Jews in the history of knowledge, such as in the Greek writings of the unknown Jewish or Samaritan author whom scholars call "Pseudo-Eupolemus," who equates Enoch with Atlas.[39] Even there, however, one might glimpse a telling tension between the impulse to laud Jewish priority in the history of knowledge and the impulse to situate the Jews within non-Jewish histories of knowledge: Pseudo-Eupolemus lauds Enoch/Atlas as the one who discovered astronomy (9.17.8), but he doubles the claim of discovery to include the father of the Jews, claiming that Abraham "discovered astronomy/astrology and the Chaldean art" (*astrologion kai Chaldaikên* [sc. *technê*] *heurein*; 9.17.3), taught the Phoenicians "the movements of the sun and moon" (9.17.5), and introduced the Egyptians "to astronomy/astrology and the rest." That this tension continues to resonate is suggested by the writings of the first-century Jewish

[37] Annette Yoshiko Reed, "The Origins of the *Book of the Watchers* as 'Apocalypse' and its Reception as 'Apocrypha,'" *Henoch* 30 (2008): 57–58; eadem, "Enoch, Eden."

[38] See esp. Jubilees 4:7, 21; 8:3–4; 10:10–14; 11:16; 12:16, 25–27; 19:14; 21:10; 45:16, on the medicinal knowledge transmitted from Noah to Shem in the broader context of a body of written and oral traditions about the stars, calendar, laws, festivals, agriculture, etc., transmitted in Hebrew in a line from Enoch, Noah, Abraham, Jacob, and Levi, to their heirs—here articulated in contrast to the non-Hebrew knowledge of celestial phenomena connected to the fallen angels and divination. Comparable but much less developed is the *Book of Watchers'* earlier appeal to cosmological knowledge mediated by Enoch in contrast to knowledge about metals, mining, cosmetics, celestial auguries, root-cutting, etc., associated with the fallen angels (1 Enoch 6–16, esp. 7:1, 8:1–3); see further Himmelfarb, *Ascent to Heaven*; Annette Yoshiko Reed, *Fallen Angels and the History of Judaism and Christianity: The Reception of Enochic Literature* (Cambridge: Cambridge University Press, 2005), 24–121.

[39] Eusebius, *Praep. ev.* 9.17.3–9; Reed, "Abraham as Chaldean Scientist," 136-142.

historian Flavius Josephus (ca. 37–100 CE), who wrote in Greek under Roman patronage and drew upon earlier traditions of the sort attested by Pseudo-Eupolemus.[40] In Josephus' presentation in the *Antiquities* (1.69–70, 154-168), Abraham is the father of Jews and inventor of monotheism, but remains meaningfully "Chaldean" in one important sense—as the astronomer from Ur, lauded by Berossus, who taught astronomy and mathematics to Egyptians.[41]

Similar traditions appear at the tail end of Late Antiquity, with similar tensions. For the author of Jubilees in the second century BCE,[42] and again for the author of *Sefer Asaf ha-Rofe* around the eighth or ninth century CE, medicine originates with angelic revelations to Noah.[43] Yet, whereas Jubilees stressed its transmission solely to Shem, and hence the Jews, *Sefer Asaf* explores the cross-cultural implications:

[40] Sterling, *Historiography and Self-Definition*, 282-284.

[41] The tension, in fact, is perhaps exemplified by the very term. Greek, Latin, Aramaic, and Hebrew terms for "Chaldean" can denote a Babylonian priest but also an astrologer of any ethnicity. The culturally-defined framing of astrological knowledge, however, stands in contrast to the astral sciences of the Hellenistic era, which was not solely Babylonian, but rather the product of a new fusion of Babylonian, Egyptian, and Greek elements. See further Reed, "Abraham as Chaldean Scientist"; Reimund Leicht, *Astrologumena Judaica: Untersuchungen zur Geschichte der astrologischen Literatur der Juden* (TSMJ 21; Tübingen: Mohr Siebeck, 2006), 11–17.

[42] Contrast the association of root-cutting, etc., with sorcery and other corrupting teachings of the fallen angels in the *Book of Watchers* (esp. *1 Enoch* 7:1; 8:3). This range of attitudes, notably, resonates with the ambivalence towards medicine within ancient Greek literature as well, wherein this domain of knowledge and expertise is often placed at the charged margins of the very category of *technê*, at its intersection with *mageia*, etc.; Serafina Cuomo, *Technology and Culture in Greek and Roman Antiquity* (Cambridge: Cambridge University Press, 2007), 7–40. That the tracing of the origins of medicine, astrology, metallurgy, etc., to good and bad angels in Jewish writings like the *Book of Watchers* and Jubilees also fits within a broader Mediterranean and Near Eastern context (i.e., whereby the origins of ambivalent *technê* could be connected to semi-divine or intermediate figures; e.g., *daimones, dactyls*) is demonstrated by Fritz Graf, "Mythical Production: Aspects of Myth and Technology in Antiquity" in *From Myth to Reason? Studies in the Development of Greek Thought* (ed. Richard Buxton; New York: Oxford University Press, 1999), 322-328.

[43] For the possibility of some connection between them, see Martha Himmelfarb, "Some Echoes of Jubilees in Medieval Hebrew Literature," in *Tracing the Threads: Studies in the Vitality of Jewish Pseudepigrapha* (ed. J. C. Reeves; Atlanta: Scholars Press, 1994), 127-136; for theories about a lost "Book of Noah," see now Michael E. Stone, Aryeh Amihay, and Vered Hillel (eds.) *Noah and His Book(s)* (SBLEJL28; Atlanta: SBL, 2011).

it asserts that the Jews are those through whom medicine came to Indians, Greeks, Egyptians, and Mesopotamians, and its list of famous physicians includes Greeks like Hippocrates and Galen, alongside Jews like Jonathan ben Zavda, Judah ha-Yarhoni, and Asaf himself.[44] Shortly after *Sefer Asaf*, moreover, one finds positions on the Jewishness of other sciences similar in concern to the *Astronomical Book* and similar in orientation to Jubilees: in *Pirqe de Rabbi Eliezer* (ch. 6), for instance, calendrical astronomy is described as a secret handed down in a priestly line from Adam to Moses, geographically-bound to Israel.

In both ancient and medieval literature, Jewish histories of knowledge are also articulated with appeal to figures who belong more unequivocally to historical time and Jewish peoplehood. Biblical claims about the scope and influence of the knowledge of Solomon (1 Kings 4:29–34), for instance, were redeployed already by the first century CE to speak to Hellenistic ideals of knowledge about the cosmos. Writing in Greek, the author of the *Wisdom of Solomon* famously makes the following claims in the name of the Israelite king:

> For it is [God] who gave me unerring knowledge of what exists,
> to know the structure of the world and the activity of the elements;
> the beginning and end and middle of times,
> the alternations of the solstices and the changes of the seasons,
> the cycles of the year and the constellations of the stars,
> the natures of animals and the tempers of wild beasts,
> the powers of spirits and the reasonings of men,
> the varieties of plants and the virtues of roots;
> I learned both what is secret and what is manifest.
> For Wisdom, the fashioner of all things, taught me. (*Wisdom of Solomon* 7:17–22; RSV)

[44] The names in *Sefer Asaf*'s list of Jewish physicians are not elsewhere attested, although some MSS identify Asaf himself with the mysterious Asaf ben Berechiah of 1 Chronicles 15:17. Shlomo Pines "The Oath of Asaph the Physician and Yohanan Ben Zabda," *Proceedings of the Israel Academy of Sciences and Humanities* 9 (1975): 223-264; Elinor Lieber, "Asaf's Book of Medicines: A Hebrew Encyclopedia of Greek and Jewish Magic, Possibly Compiled on an Indian Model," *Dumbarton Oaks Papers* 38 (1984): 233-249; Reed, "Origins."

Centuries later, when writing of the relationship of Judaism and philosophy in Judeo-Arabic, Judah Halevi (ca. 1075–1141) similarly appeals to Solomon to claim all sciences as originally "Jewish":

> Did he not, with the assistance of divine, intellectual, and natural power, converse on all sciences [*'ulūm*]? The inhabitants of the earth traveled to him, in order to carry forth his learning, even as far as India. Now the roots and principles of all sciences were handed down from us [i.e., the Jews] first to the Chaldaeans, then to the Persians and Medians, then to Greece, and finally to the Romans.[45]

Whether or not the latter knew the former,[46] the parallels remain significant. Both the *Wisdom of Solomon* and *Kitab al Khazari* use the famously wise and cosmopolitan king to evoke a vision of Jewish wisdom as encompassing the totality of scientific knowledge current, valued, and sought among non-Jews of their times. If the first-century author of the *Wisdom of Solomon* did so to counter the totalizing claims and intellectual prestige of Hellenistic *paideia* in early Roman Egypt,[47] Judah Halevi answers much the same problem, many centuries later, in the context of the recovery and cultivation of Hellenistic texts and traditions by Islamic intellectuals.[48]

Perhaps telling, however, is the caveat with which Judah Halevi here concludes: "On account of the length of this period [of transmission], and the many disturbing circumstances, it was forgotten that they had originated with the Hebrews" (*Kitab al Khazari* II 66). Even as he proclaims the Jewishness of "all sciences," he also

[45] I am grateful to Nicholas Harris for consulting the Judeo-Arabic of this passage. Judah Halevi, *Kitab al Khazari* (trans. Hartwig Hirschfeld; London: Routledge, 1905), II 66, 124.
[46] Interestingly, Nahmanides (ca. 1194–1270) knew an Aramaic version of the *Wisdom of Solomon* and quotes the above-cited passage in the preface to his commentary to the Torah/Pentateuch, see Alexander Marx, "An Aramaic Fragment of the Wisdom of Solomon," *JBL* 40 (1921): 58-60.
[47] Cf. John J. Collins, "Cosmos and Salvation: Jewish Wisdom and Apocalyptic in the Hellenistic Age," *History of Religions* 17.2 (1977): 121-142.
[48] Cf. Langermann, "Science and the Kuzari"; Abdelhamid I. Sabra, "The Appropriation and Subsequent Naturalization of Greek Science in Medieval Islam," *History of Science* 25 (1987): 225-243; Walbridge, "Explaining Away the Greek Gods"; Gutas, *Greek Thought*.

admits that this ancient history is not acknowledged by the very peoples among whom this knowledge flourishes in his own times.[49] In this caveat, we may glimpse something of what is at stake in the assertion of the antiquity of Jewish contribution to scientific knowledge for some earlier authors as well. In *contra Apionem*, for instance, Josephus complained that Apion charged that the Jews "have not produced remarkable men, such as inventors [*heurêtas*] of some *technai*" (2.135) and that Apollonius Molon called the Jews "the most untalented of barbarians and ... the only ones to have contributed no invention [*heurêma*] of use to human life" (2.148).[50] If Louis Feldman is correct, his characterization of Abraham as astronomer in the *Antiquities* may have been shaped, at least in part, by the concern to counter precisely such accusations.[51]

In the first century and the twelfth—even if not necessarily always before or in between—assertions of Jewish primacy or participation in cross-cultural endeavors of learning like astronomy, mathematics, or medicine could serve to answer accusations against the Jews for an allegedly misanthropic isolation from the cosmopolitan endeavor of furthering human knowledge. This concern, however, makes it all the more poignant that such challenges could simultaneously serve as opportunities for reflection on the proper scope and bounds of Jewish inquiry and learning. After all, for

[49] Notably, Halevi does assert its continued cultivation within Judaism, albeit in service to the distinctively Jewish domain of halakhah (e.g., *Kitab al Khazari* II 64; III 41; IV 25). He stresses, e.g., that "our Sages were, without doubt, acquainted with the revolutions of the sun and the other planets," inasmuch as "all branches of science were required for the practice of Jewish Law" (*Kitab al Khazari* II 64). See further Langermann, "Science and the Kuzari," 495–522.

[50] Even as Josephus answers such accusations in part through his portraits of Abraham and Moses, he also makes the argument that it is the very piety [*eusebeia*] of the Jews, their unwavering observance of their ancestral customs, that lies at "the origin of the charge that some have raised against us, that we have produced no inventors [*heurêtas*] of novel deeds or words" (*contra Apionem* 2.182). Although Josephus is certainly not the last Jewish thinker to contrast piety, as the domain of the Jews, with areas of interest and inquiry associated for the sake of contrast with the Greeks, this is just one of the ancient Jewish approaches to organizing knowledge—or so I attempt to suggest in Reed, "Was there Science in Ancient Judaism?"

[51] Louis Feldman, *Josephus's Interpretation of the Bible* (Berkeley: University of California Press, 1998), 223-289.

Josephus, as for Judah Halevi, the claim of the antiquity of Jewish involvement in sciences draws its power from the perception of certain types of knowledge about the cosmos, the human body, and the stars as valuable across the bounds of different cultures, traveling between them without diluting or corrupting the ancestral customs or distinctive cultural identities of any of them.[52] By such accounts, to be a Jewish astronomer like Enoch or Abraham, or a Jewish doctor like Noah, Shem, or Asaf is to be no less "Jewish," even if it is also to master domains highly valued beyond the bounds of Judaism.

These various premodern attempts to insert Jews into the history of sciences, and "science" into the history of Judaism, thus share an important set of assumptions with the Greek, Roman, and Islamic approaches noted above. In all of them, disciplines such as astronomy, cosmology, mathematics, and/or medicine are deemed forms of knowledge eminently *translatable* across different times and cultures—whether by virtue of connection to the visible world or human body, or to shared theoretical models or observational methods for collecting, preserving, and systematizing cross-culturally accumulated bodies of knowledge about them.[53] In ideological import and cross-cultural prestige, their ancient forms can be heuristically likened to the disciplines now arrayed under our modern category of "science" (even if constructs of that sort must remain ever-shifting, e.g., with antecedents in antiquity sometimes including astrology, alchemy, physiognomics, etc., while antecedents in the Middle Ages sometimes

[52] Such a view is striking precisely because other types of knowledge are clearly perceived as corrupting when adopted from one culture into another. Paradigmatic in the case of ancient Judaism is knowledge about the divine, as richly demonstrated by Mark S. Smith, *God in Translation: Deities in Cross-Cultural Discourse in the Biblical World* (FAT 57; Tübingen: Mohr Siebeck, 2008). Note also, however, Greek and Roman attitudes towards various kinds of knowledge marginalized as *mageia/magia*, often with appeal to purported foreignness, in a manner akin to the Jewish discourse about "the nations," "the ways of the Amorite," etc.; see esp. Pliny, *Natural History* 30; Graf, "Mythical Production," 20–60; Veltri, "The Rabbis and Pliny the Elder". On Greek ideas about "ancestral customs" [*ta patria ethê*] and cultural difference, see now Benjamin Isaac, *The Invention of Racism in Classical Antiquity* (Princeton: Princeton University Press, 2006).

[53] My use of the category of "translatability" here extends the model outlined in Smith, *God in Translation*, with respect to discourses about divinity.

excluding those types of empiricism so strongly associated with "the scientific method" today).[54] In this sense, then, the notion of "ancient Jewish sciences" may prove interesting for the history of knowledge precisely because it seems a bit paradoxical. What might appear as a paradox is perhaps, rather, a productive tension, which finds expression in the range of ways in which ancient and medieval Jews engaged knowledge simultaneously within and beyond Judaism.

2. Rethinking "religion" and "science"

I have dwelt in some length upon premodern perspectives on knowledge and cultural specificity in part because similar concerns still resonate today. Running through much of the scholarship on "ancient Jewish sciences" are questions not so dissimilar to those that seem to have troubled Josephus and Judah Halevi. Did ancient Jews contribute anything to the progress of human knowledge in their time? Or was ancient Jewish engagement with ancient sciences largely limited to "borrowings" from Babylonian or Hellenistic traditions? Do Jewish sources of the sort that one finds at Qumran have any place in scholarship on the history of science?

Of course, scholars of ancient China and Mesopotamia have long faced similar challenges arguing for a place in a history of science still commonly presumed to be a story about Greeks in antiquity and Europeans in modernity.[55] Yet, just as popular surveys of the history of science sometimes read like a litany of discoveries leading to the modern West, as if extending the line of ancient Greek heurematography,[56] so interdisciplinary efforts at the inclusion of other peoples can run the same risks as the ancient Jewish responses noted above: some arguments for inclusion can unintentionally re-inscribe the assumption that "science" is a stable category with a

[54] Instructive is the example of shifting perspectives on alchemy, on which see Lawrence M. Principe, "Alchemy Restored," *Isis* 102.2 (2011): 305-312.
[55] Cf. Geoffrey E. R. Lloyd, *The Ambitions of Curiosity: Understanding the World in Ancient Greece and China* (Cambridge: Cambridge University Press, 2002); Francesca Rochberg, "A Consideration of Babylonian Astronomy within the Historiography of Science," *Studies in the History and Philosophy of Science Part A*. 33 (2002): 661-684; eadem, *The Heavenly Writing: Divination, Horoscopy, and Astronomy in Mesopotamian Culture* (Cambridge: Cambridge University Press, 2004).
[56] Cf. Zhmud, *Origin of the History of Science*, 1–22.

singular lineage, inevitable in its progress towards the politically-dominant power of the present.[57]

An essentialist ideal of "science" formed part of the vision of the history of science promoted by founding figures in the field, who popularized an enduring narrative of Greek/European exemplarity, even as they stressed the uniquely transcultural and transhistorical character of their topic.[58] Especially since the 1960s, this vision has been widely critiqued for "essentialist universalism," "Whig history," and Eurocentrism, as well as for the misleading projection of continuity in the "Western" production and transmission of

[57] This is perhaps best exemplified by the discussion surrounding the "Needham question," whereby the sciences and technologies of Chinese and other non-"Western" cultures have been studied through the lens of the question of *why* they were not home to the Scientific Revolution that occurred in 19th-century Europe. The question itself presumes—and thus reinforces—a model of the history of sciences as a singular line of inevitable progress towards the Western present. Cf. Joseph Needham et al., *Science and Civilisation in China*, 7 vols. (Cambridge: Cambridge University Press, 1954-2004); Lloyd, *The Ambitions of Curiosity;* Andrew Brennan, "The Birth of Modern Science: Culture, Mentalities and Scientific Innovation," *Studies in History and Philosophy of Science* 35 (2004): 199–225. On the ways in which assumptions about "progress" have similarly shaped scholarship on ancient *technai*, see Cuomo, *Technology and Culture*, 3–4.

[58] On Paul Tannery, George Sarton, and Alexandre Koyré, in particular, see further Lewis Pyenson, "The ideology of Western Rationality: History of Science and the European Civilizing Mission." *Science & Education* 2.4 (1993), 329-343. Often cited in this regard are Sarton's assertions that "science" is "the only human activity which is truly cumulative and progressive," as well as his promotion of its history as the record of a singular march of progress in "the acquisition and systematization of positive knowledge"; e.g., Sarton, *Introduction to the History of Science*, vol. 1: *From Homer to Omar Khayyam* (Baltimore: Williams and Wilkins, 1929), 4. Yet, as Elshakry ("When Science Became Western," 107) notes, "historians of science like Sarton (and Joseph Needham) were driven by the desire to demonstrate the ancient and medieval or early modern contributions of Eastern civilizations. But once the narrative of the rise of Western science was set in place, other counternarratives were implied, with their distinctive vocabulary of stagnation, decline, and dark ages"; it was the invention of the idea of the "Scientific Revolution" in the 1930s that decisively "sealed off the West from the rest" and "helped to set the agenda for how the discipline itself would subsequently view the world, as a new emphasis on a universal and unilinear history of science merged seamlessly with postwar modernization theories." It is interesting to note in this regard that Sarton's *Introduction to the History of Science* was notable in its own time for its inclusion of Jews; see further Joshua Finkel, "Review: Sarton on the History of Science," *JQR* 18.4 (1928): 445-448.

knowledge.[59] Nevertheless, it remains widely diffused due to its integration into textbooks and teaching,[60] and it is often reinforced by the criteria chosen to defend the scienticity of ancient and/or non-European traditions in particular. Speculation about possible motives for interest and inquiry into the visible world, for instance, frequently privilege disinterest as a mark of the scientific character of knowledge, consistent with modern ideals of "science" as knowledge pursued "for its own sake," especially in contrast to knowledge in the service of religious aims. Similarly, discussions of methods of inquiry often argue for the scienticity of practices with some basis in observation that can be likened to the empiricism and experimentalism of the modern "scientific method." When assessing the results, moreover, it can be tempting to celebrate as "science" those developments that seem to make "progress" towards our own present.

What is significant, for our purposes, is the double challenge thus posed for the emergence of "ancient Jewish sciences" as an area of study. Recent historiographical shifts have resulted in divergent assumptions about what even constitutes "science" among historians of science, on the one hand, and historians of Judaism, on the other. Among the former, it is largely taken for granted that "science" is a modern construct. Yet among the latter—as in the popular press and public imagination—older notions of "science" and "the West" remain

[59] The critique of early scholarship in the history of science for presentist biases was popularized by Thomas Kuhn, *The Structure of Scientific Revolutions*, 3rd ed. (Chicago: University of Chicago Press, 1996 [1962]), see esp. pp. 1–3, 137-141. Relevant for our purposes is his emphasis on "the persistent tendency to make the history of science look linear or cumulative" by reinterpreting the past to fit the epistemological values and assumptions of the present: "Partly by selectivity, and partly by distortion, the scientists of earlier ages are implicitly represented as having worked upon the same set of fixed problems" (p. 138).

[60] Consistent with the efforts of scholars like Sarton to promote the history of science as an integral part of a Humanistic education, such older views continue to shape ideas about the history of science in the popular imagination and, as a result, to have an enduring impact on interdisciplinary inquiries into ancient and/or non-Western sciences. On the success of the founding figures in the field in integrating these narratives into science textbooks, high-school and college courses on the "history of Western civilization," see Kuhn, *The Structure*, 137-140; Roberts, "Situating Science in Global History."

widespread. We shall return below to consider some of the challenges posed to the study of "ancient Jewish sciences" by the destabilization of the category of "science" among historians of science and by the emphasis on the modern European contexts of its construction. First, however, it may be helpful to delve in a bit more detail into the challenges of defending the topic to what has so far been the main presumed audience for studies of "ancient Jewish sciences," namely, historians of Judaism.

Particularly insofar as much of the relevant data for "ancient Jewish sciences" concern calendrical, astrological, and physiognomic practices widespread in antiquity but not perceived as "scientific" today, recent attempts to make a place for the study of "ancient Jewish sciences" have largely centered on analogy and filiation to their Greek and Mesopotamian counterparts.[61] As perhaps to be expected at this preliminary stage of the emergence of new approaches to "ancient Jewish sciences," moreover, recent works have been faced with the need to answer past studies that had been pursued without the benefit of newer evidence or approaches. It was Otto Neugebauer's 1964 assessment of the Babylonian astronomy of the Enochic *Astronomical Book* as "primitive,"[62] for instance, that lead Michael E. Stone[63] to interpret the choice as "a deliberate act of archaism ... aris[ing] either from the conscious rejection of Greek science or else the creation of a social context into which such science did not penetrate." For Stone, thus, the astronomy of the *Astronomical Book* was not evidence for "ancient Jewish sciences" as much as an expression of an anti-scientific and ultra-traditionalist stance, likely cultivated in "separatist" circles.[64]

Similar assumptions about the linearity of scientific progress inform another important precedent for the present discussion, albeit with the opposite results. P. S. Alexander sought to argue that "the approach to nature displayed in the Enochic *Book of the Heavenly*

[61] E.g., Popović, *Reading the Human Body*, 211-212; Ben-Dov, *Head of All Years*, 3–4—both following the important precedent of Alexander, "Enoch and the Beginnings," 224.
[62] Cf. Ben-Dov, *Head of All Years*, 13.
[63] Stone, "Enoch, Levi," 251-252.
[64] Stone, "Enoch, Levi," 252; cf. Reed, *Fallen Angels*, 68–69.

Luminaries is unprecedented in Jewish literature," reflecting an anti-traditionalist stance, cultivated among Jews less interested in the Torah than in "alien wisdom."[65] To do so, he pressed for a pre-Hellenistic date for the *Astronomical Book*, prior to the developments that led Stone and others to dismiss its science as "out-dated."[66] What Stone interpreted as reflecting "an intellectual cast that was not interested in contemporary scientific knowledge,"[67] Alexander thus asserted as evidence for "a turning-point in Jewish intellectual history—the emergence, for the first time, of what might properly be called a scientific attitude."[68]

More recent discussions suggest the situation was more complex than known to Stone or Alexander at the time of their ground-breaking essays. Popović rightly stresses, for instance, how publication of the full range of relevant materials from Qumran has since shed doubt on any simple equation of dependence on Babylonian astronomy with the archaizing retention of "out-dated" views or the self-conscious rejection by separatists of some self-evidently superior Hellenistic science.[69] In addition, newer evidence and approaches—masterfully synthesized by Ben-Dov[70]—now push us to rethink the notion that any seriously "scientific" interest or engagement with Babylonian traditions must have pre-dated the Hellenistic period.

65 This conjecture, as noted below, forms part of his broader project of positing a "Enochic"/priestly tradition running alongside a dominant "Mosaic"/legal tradition within Judaism, see, e.g., Philip S. Alexander, "What Happened to the Jewish Priesthood After 70?," in *A Wandering Galilean: Essays in Honour of Seán Freyne* (ed. Zuleika Rodgers, Margaret Daly-Denton, and Anne Fitzpatrick McKinley; Leiden: Brill, 2009), 3–34; cf. Rachel Elior, *The Three Temples: On the Emergence of Jewish Mysticism* (Oxford: Littman Library, 2004); Alexander, "Enoch and the Beginnings," 240.

66 Notably, Alexander is thus among the minority of scholars who locate the *Astronomical Book* in the Persian period, rather than in the early Hellenistic period; cf. VanderKam, *Enoch and the Growth*, 83–88. Hence, even those who might agree with his assessment of its stance towards Jewish and non-Jewish knowledge may well be wary of his speculation that "[s]ometime in the late Persian period, say around 450–400 BCE, under the influence of Persian and, ultimately, of Babylonian ideas, Jews for the first time became interested in producing scientific models of the workings of the natural world" (Alexander, "Enoch and the Beginnings," 237).

67 Stone, "Enoch, Levi," 252.

68 Alexander, "Enoch and the Beginnings," 240.

69 Popović, "Physiognomic Knowledge," 223-224.

70 Ben-Dov, *Head of All Years*, 12–13, 282-287.

Traditionally, Mesopotamian influence on Judaism has been associated with the Near Eastern cultural contexts of ancient Israel and the earliest traditions in the Hebrew Bible, in contrast to the Hellenistic influence deemed determinative after the conquests of Alexander the Great. Recent insights into the *Nachleben* of cuneiform culture, Akkadian sciences, and their scribal pedagogies and curricula, however, have demonstrated the continued vitality and development of Babylonian sciences under Seleucid rule and beyond,[71] as well as their continued and dynamic impact on Judaism, possibly well into Late Antiquity.[72] As Popović rightly emphasizes, the Jewish literature of the Hellenistic age attests the influence of both Mesopotamian and Greek traditions of astronomy, cosmography, geography, and physiognomy.[73] Furthermore, Ben-Dov raises the possibility that Jews may have played some role in the circulation of scientific traditions, not merely as passive receivers of knowledge radiating from some single Greek center, but rather as part of a dynamic constellation of interconnected locales.[74]

If recent research has done much to transform our image of the place of Jews with the history of ancient sciences, it has perhaps been

[71] E.g. Markham Geller, "The Last Wedge," *ZA* 87 (1997): 43-95; Rochberg, *Heavenly Writing*.

[72] E.g., Geller, "The Last Wedge;" idem, "Babylonian Influence on Hellenistic Judaism," in *Civilizations of the Ancient Near East* (ed. J.M. Sasson; New York: Scribner, 1995), vol. I, 43-54; idem, "The survival of Babylonian Wissenschaft in Later Tradition," *Melammu Symposia* I (ed. S. Aro and R. M. Whiting; Helsinki: Helsinki University Press, 2000), 1-6; idem, "An Akkadian Vademecum in the Babylonian Talmud," in *From Athens to Jerusalem, Medicine in Hellenized Jewish Lore and in Early Christian Literature* (ed. S. Kottek, M. Horstmanshoff, G. Baader, and G. Ferngren; Rotterdam: Erasmus, 2000),13-32; Henryk Drawnel, "Priestly Education in the *Aramaic Levi Document* (Visions of Levi) and *Aramaic Astronomical Book* (4Q208–211)", *RQ* 22.4 (2006): 547-574; idem, "Between Akkadian *tupšarrutu* and Aramaic *spr*: Some Notes on the Social Context of the Early Enochic Literature," *RQ* 24 (2010): 373-403; Ben-Dov, *Head of All Years*; idem, "Scientific Writings in Aramaic and Hebrew at Qumran: Translation and Concealment," in *Aramaic Qumranica: The Aix-en Provence Colloquium on the Aramaic Dead Sea Scrolls* (STDJ 94; ed. K. Berthelot and D. Stökl Ben Ezra; Leiden: Brill, 2010), 379-402.

[73] Mladen Popović, "The Emergence of Aramaic and Hebrew Scholarly Texts: Transmission and Translation of Alien Wisdom" in *The Dead Sea Scrolls: Transmission of Traditions and Production of Texts* (STDJ 92; ed. S. Metso, H. Najman, and E. Schuller; Leiden: Brill, 2010), 84.

[74] Ben-Dov, *Head of All Years*, 265-266.

trickier to move beyond older assumptions about "science" and ancient Judaism. In the above-noted essay,[75] I noted a tendency to define "Judaism" in terms of a modern characterization of "religion" (esp., ethics, halakha) and, thus, to interpret evidence for Jewish engagement with sciences as cases of "foreign" influence or examples of the engagement of individual Jews in the knowledge-enterprises of their non-Jewish rulers and neighbors. What is assumed and asserted by such a move—I suggested—is an anachronistic understanding of "religion" and "science" as self-contained and mutually-exclusive approaches to explaining the world and human experience. Inasmuch as ancient Judaism is thereby reduced to the religious, much can be lost in the process. To neglect of the Jewishness of Jewish engagement with ancient sciences is to skew our understanding of the richness of reflection on the stars, cosmos, and human body within the history of Judaism.

Here, I would like to highlight a related scholarly tendency with similarly wide ripples of ramifications, namely, the tendency to overlay the modern dichotomy of "religion" vs. "science" upon other dichotomies common in the modern historiography of ancient Judaism, including traditional contrasts like Semitic vs. Greek, Near Eastern vs. Hellenistic, and Jewish vs. foreign, but also newer ones like "Mosaic" vs. "Enochic."[76] To understand the logic behind this move, as well as its puzzlingly perennial appeal, we might note a broader pattern in the study of the ancient Near East. In a recent study of Mesopotamian sciences, for instance, Francesca Rochberg points to a long-standing tendency to dismiss Near Eastern knowledge-enterprises as not truly "scientific" by virtue of the "religious" motives imputed to their inquiries:

> ... a clear distinction between science and religion, and therefore also between knowledge and belief, was an important device in the defining of science by the 1960s. The opposition rendered between reason and scientific

[75] Reed, "Was there Science in Ancient Judaism?" 461-467.

[76] Gabriele Boccaccini, *Beyond the Essene Hypothesis: The Parting of the Ways between Qumran and Enochic Judaism* (Grand Rapids, Mi.: Eerdmans, 1998); idem, *Roots of Rabbinic Judaism: An Intellectual History, From Ezekiel to Daniel* (Grand Rapids, Mi.: Eerdmans, 2002).

> knowledge on one hand and religion, superstition, and unscientific belief on the other informed a historiography that saw the necessity of a break with some religious or mythological tradition ... before the "birth" of science was possible ... This view evoked not only an Enlightenment sensibility but also a neoevolutionist cognitive anthropology, as Near Eastern forms of inquiry into natural phenomena were deemed necessarily more primitive than those of the Greeks.[77]

Modern notions of the timeless and essential conflict between "science" and "religion"—Rochberg here suggests—were not only paired with older notions about the cognitive differences between Greek/"Western" and Near-Eastern/"Oriental" peoples, but they served as one means of construing "science" as a primarily "Western" phenomenon and as exemplar of the Greek exceptionalism to which modern Europe is deemed as heir:

> Despite the acknowledgment of an intellectual transmission from Babylonia to the Greeks, when it came to general histories of science, Babylonian learning (along with that of other non-Greek ancient sources such as those from Egypt, India, and China) would be contrasted with Greek "knowledge" in one of two ways. What the eastern ancients "knew" was categorized either as mere craft ... or as theological speculation not anchored by logical, causal, or rational inquiry into physical phenomena. [78]

Hence, for instance, scholars such as E. H. Hutton could posit that "the philosophers of the Ionian school combined theorizing about the universe with knowing some facts and this made their work so unique ... Eastern 'sages' too were speculating about the world, but they were guided by religious and moral feelings rather than by the desire to understand external reality ... [and] thus the Orientals never developed science."[79]

[77] Rochberg, *Heavenly Writing*, 18.
[78] Rochberg, *Heavenly Writing*, 16.
[79] Ibid, 16-18.

What is interesting, for our purposes, are the echoes of such ideas even in relatively recent research on ancient Judaism. In the above-noted essay by Alexander, for instance, the contrast between "religion" and "science" is embraced in precisely such terms. What Alexander ultimately wishes to argue, in fact, is that early Enochic interest in sciences is akin to "the rise of the Ionian school of Greek philosophy and science"—a cognitive shift resulting in "a view of nature which ... was radically new and which can for the first time be meaningfully labeled as 'scientific'" precisely because of an alleged "break from pre-existing mythical and epic pictures of the world."[80]

It is only with some hermeneutical gymnastics, of course, that the early Enochic literature can be presented as an exemplar of a break with the mythical. To do so, Alexander must read these ancient sources against the grain, dismissing their appeal to Enoch as simply a ruse. He downplays the richness of the Mesopotamian roots and matrix of both the Hebrew Bible and the early Enochic literature.[81] Partly as a result, Alexander reduces the latter's appeal to Enoch as a "recourse" to pseudepigraphy to conceal the true motives of the authors, which—he speculates—was "to domesticate within Jewish tradition a body of alien wisdom ... fully aware of the newness of their doctrine—that they were propagating ideas never before heard in Israel."[82]

This argument forms part of Alexander's broader project to posit a bifurcation in Jewish intellectual history between a putative Enochic/"priestly" paradigm, producing "scientific" and "mystical" traditions, and the more familiar Mosaic paradigm associated with the Torah and its Rabbinic interpreters.[83] The problems with approaches of this sort are well known, not least because similar theories[84] have met with intensive criticism from specialists in every field with which they

80 Alexander, "Enoch and the Beginnings," 230-236.
81 Cf. VanderKam, *Enoch and the Growth*; Helge S. Kvanvig, *Roots of Apocalyptic: The Mesopotamian Background of the Enoch Figure and of the Son of Man* (WMANT 61; Neukirchen-Vluyn: Neukirchener Verlag, 1988).
82 Alexander, "Enoch and the Beginnings," 232. See VanderKam, in this volume.
83 See now Alexander, "Jewish Priesthood."
84 E.g., Boccaccini, *Beyond the Essene Hypothesis*; idem, *Roots of Rabbinic Judaism*; Elior, *The Three Temples*.

intersect (e.g., Enochic literature, Dead Sea Scrolls, Jewish calendar, Rabbinics, Hekhalot literature).[85] For our present purposes, it suffices to note the degree to which Alexander draws upon the modern contrast between "religion" and "science" to ground the plausibility of a dichotomous understanding of ancient Judaism as split between "Enochic" and "Mosaic" paradigms, as in the following assertion:

> The circles which stand behind the Books of Enoch were, I would argue, proposing an Enochic paradigm for Judaism in opposition to the emerging Mosaic paradigm —a paradigm based primarily on science as opposed to one based primarily on law. They were innovators: they had taken on board some of the scientific thought of their day and had used it aggressively to promote a new Jewish worldview.[86]

The claim of a distinction between "Enochic" and "Mosaic" worldviews, in other words, is here tied to the contrast between "science," on the one hand, and law and myth, on the other—with Jewish interest in the former further associated with an embrace of "alien wisdom" and a radically innovative break from Torah-centered Judaism.

[85] See further, e.g., Sacha Stern, "Rachel Elior on Ancient Jewish Calendars: A Critique," *Aleph* 5 (2005): 287-292; Martha Himmelfarb, "Merkavah Mysticism since Scholem: Rachel Elior's The Three Temples," in *Wege mystischer Gotteserfahrung: Judentum, Christentum und Islam* (ed. Peter Schäfer; Munich: Oldenbourg, 2006), 19–36; John J. Collins, "'Enochic Judaism' and the Sect of the Dead Sea Scrolls," in *The Early Enoch Literature* (JSJSup 121; ed. Gabriele Boccaccini and John J. Collins; Leiden: Brill, 2007), 283-300; James C. VanderKam, "Mapping Second Temple Judaism," in *The Early Enoch Literature*, 1-20; Peter Schäfer, *The Origins of Jewish Mysticism* (Tübingen: Mohr Siebeck, 2009); Ra'anan S. Boustan, "Rabbinization and the Making of Early Jewish Mysticism," *JQR* 101.4 (2011): 482–501. As noted below, "science" also plays a part in Elior, *The Three Temples*; her argument about the emergence of Enochic/"priestly" circles is largely based on a claimed distinction between solar-calendar supporters and lunar-calendar supporters in Second Temple times. Stern's detailed critique of her position in "Rachel Elior on Ancient Jewish Sciences" is predicated on a misunderstanding or misrepresentation of the evidence stands as a "parade example" of the importance of research on "ancient Jewish sciences"—and the study of calendrical astronomy in particular—for the historiography of Judaism more broadly.

[86] Alexander, "Jewish Priesthood," 234.

To be sure, Alexander frames his 2002 essay as a speculative foray, and his assertions are largely un-referenced, and explicitly experimental and exploratory in tone.[87] Yet, for our present purposes, it provides an interesting example to illustrate some of the ways in which modern assumptions about "science" and "religion" can inform the interpretation of ancient Jewish texts and history, making explicit some of the assumptions possibly implicit in other discussions as well.[88] In addition, as an early example of a forceful call to see Second Temple Jewish evidence as important for the history of science, Alexander's 2002 essay has been widely cited in recent research on "ancient Jewish sciences."

A recent article by Popović, for instance, articulates the significance of texts such as 4Q186, 4Q317, 4Q318, and 4Q561 by means of comparison and contrast with "scientific interests evinced by *1 Enoch* and the lists of revealed things in apocalyptic texts," in order to "extend Alexander's discussion of ancient Jewish science by taking

[87] Notably too, I take issue here with only one portion of what is a richer and broader discussion (i.e., the portion on early Enochic materials, esp. Alexander, "Enoch and the Beginnings," 230-236). Even if some of his speculations have not survived further scrutiny, moreover, many of them have (e.g., his insights about language choice, as extended by Ben-Dov, on which see below), and many should be followed up further (e.g., his assessment of Rabbinic attitudes).

[88] That "ancient Jewish sciences" can have consequences for the historiography of ancient Judaism is also clear from the similarly dichotomous model posited by Elior, *The Three Temples,* which pivots on a neatly schematic but largely unfounded contrast between solar and lunar calendars (cf. Stern, "Rachel Elior on Ancient Jewish Sciences"; Himmelfarb, "Merkavah Mysticism,"25–29; Ben-Dov, *Head of All Years,* 4–5). Elior interprets early Enochic materials as attesting the origins of a visionary and sectarian "mystical" movement in early priestly defenders of a schematic solar calendar, purportedly in resistance to the proto-Rabbinic defenders of a more practically-oriented lunar calendar. Based on this contrast, she characterizes what she calls the "secessionist priesthood" as committed to "[t]he mathematization of the universe and its manifestations in the cycles of nature and the cycles of sacred service" in contrast to the proto-Rabbinic and Rabbinic factions "who refused to subjugate time and its divisions to an eternal, unchanging divine order" (Elior, *Three Temples,* 213). Although Elior (*Three Temples,* 212) claims to find in the sources "a sharp polar relationship between that [priestly] literature and rabbinic positions—an antithetical correlation, with one corpus negating what was advocated by the other," her theory has been widely critiqued for misrepresenting the very texts that she cites to support it; see esp. Himmelfarb, "Merkavah Mysticism."

into account some of the texts from the Dead Sea Scrolls corpus."[89] Following Alexander, Popović describes the varying degrees of scientific engagement attested in the Dead Sea Scrolls as "adaptations and emulations of alien wisdom" that can be contrasted with the "Hebrew wisdom" of the Hebrew Bible and the *Wisdom of Ben Sira*.[90] Alexander's characterization of the motives behind the early Enochic materials also provides the basis for his contrast of these materials with other evidence for Jewish engagement with ancient sciences discovered among the Dead Sea scrolls. Just as Alexander interprets the appeal to Enoch as a ruse to "domesticate" and justify the "alien wisdom" of Babylonian astronomy in biblical, Jewish, and "religious" terms,[91] so Popović posits for 4Q186, 4Q317, 4Q318, and 4Q561 that "the apparent lack of an attribution to a pseudepigraphical figure as an authoritative voice ... indicates [that their] scientific interests did not need such justification."[92] Inasmuch as the latter seem to have lacked the "Enochic interest in divine, eschatological judgment" as well as "any scriptural exemplar," moreover, Popović suggests that they may

[89] Popović, "The Emergence," 82, 84–85.
[90] Ibid, 83, 114. Even aside from the question of the degree to which one can take ben Sira's traditionalist claims at face value, it remains that the contrast emblematizes the selective anachronism of the scholarly discussion. As Ben Wright, Seth Sanders, and others noted at the ISAW conference on which this volume is based, the *Wisdom of Ben Sira* and other sapiential writings provide perhaps the strongest precedents with Judaism for empiricism and observation-based inquiry; even in its earliest articulations, the Jewish Wisdom tradition is characterized by an emphasis on experience as a source of knowledge. To treat the *Wisdom of Ben Sira* only as a point of contrast with Enochic and other apocalypses vis-à-vis "science" is thus misleading in multiple ways, distracting from a potentially rich source for understanding the history of Jewish approaches to knowledge about the cosmos, etc., while also using the quotation of selected passages to suggest a larger contrast with Enochic materials than careful analysis of the entire works reveal. See now Benjamin Wright, "1 Enoch and Ben Sira: Wisdom and Apocalyptic in Relationship," in *The Early Enoch Literature*, 159-179, and further references there.
[91] Alexander, "Enoch and the Beginnings," 232.
[92] Popović, "The Emergence," 97–98. To place such emphasis on what is lacking in these texts to characterize their motives, etc., proves a bit tenuous given their fragmentary state of preservation. See below, however, further to the importance of Popović's point concerning form and framing.

attest a category of writings otherwise unknown in ancient Judaism, namely, what he terms "purely scientific texts."[93]

Although Popović repeatedly signals the dangers of anachronism in applying terms like "religious," "scientific," and "secular" to ancient cultures, he nevertheless maintains these terms as heuristic for analyzing the physiognomic materials from Qumran and arguing for their character as "science."[94] When he argues that "physiognomic and physiognomic-astrological lists from Qumran ... represent forms of ancient Jewish science," for instance, he stresses that these "lists from Qumran were possibly not framed by religious interests" and, as such, could attest "a well-educated body of people in ancient Jewish society ... both priestly and secular scribes or scholars ... interested in contemporary scientific learning, not just in 'outdated' forms of Mesopotamian astronomy as in the *Astronomical Book*."[95] Despite his own caution and concerns, his application of these terms may unintentionally reinforce older essentialist assumptions about the mutual exclusivity of "religion"/Judaism and "science"/secularity—in part, because of the conceptual baggage carried by the very distinction, and in part, by virtue of his dependence on Alexander in applying this distinction dichotomously to ancient Jewish materials.[96] Inasmuch as this distinction is used to draw out the differences between "scientific material that has been framed or reworked into other writings such as apocalyptic texts and ... those Dead Sea Scrolls from Qumran that provide manuscript evidence for 'actual' scholarly/scientific texts," it may distract from an argument that may prove critical—as we shall see—for mapping new approaches to the transmission of scientific knowledge within Judaism.[97]

In comparison with Popović, Ben-Dov argues for understanding "ancient Jewish sciences" both as more "scientific" and as more "Jewish." He deals with much the same range of evidence, albeit defining "ancient Jewish sciences" a bit more broadly to include

[93] Ibid, 86, 87.
[94] Popović, *Reading the Human Body*, 15, 222; idem, "The Emergence," 82–83, 86.
[95] Popović, "The Emergence," 82, 84–85.
[96] Alexander, "Enoch and the Beginnings," 230-236.
[97] Popović, "The Emergence," 83.

astronomy, astrology, geography, metrology, physiognomy, and exorcism.[98] Programmatic for his 2008 monograph, for instance, is his aim to "demonstrate that the Jewish scientific tradition ... constituted an integral part of the astronomical knowledge current in the Ancient Near East during the Persian and early Hellenistic periods" but also that "the emulation of this knowledge in Jewish circles lead to a new synthesis, perceptibly different from the main streams of astronomical teaching existent in Babylonia, Greece, Egypt, and India ... a self-contained intellectual construct."[99] What he suggests, in other words, is the coexistence of a range of practices of "ancient Jewish sciences" akin the continuum we have seen in ancient and medieval representations of them—with "sciences" as a crossing or meeting-point between Jewish and other cultures, akin in spirit to that imagined by Pseudo-Eupolemus and Josephus, on the one hand, and as a distinctively Jewish domain of knowledge, akin in spirit to what is imagined in Jubilees and *Pirqe de Rabbi Eliezer*, on the other.

For Ben-Dov too, the comparison of Enochic and other Qumran materials proves pivotal. Whereas Popović focuses largely on differences in literary forms and settings, Ben-Dov stresses differences in language.[100] That so much of the earliest known scientific literature is attested in Aramaic provides a key for him to solving multiple puzzles pertaining to Jewish engagement in cross-cultural scientific endeavors, and to the transmission of scientific knowledge within and beyond Judaism. With the aid of the evidence from Qumran, Ben-Dov mounts a persuasive case that the Aramaic language served as a vehicle for the transmission and cross-cultural diffusion of Mesopotamian traditions known to us primarily in cuneiform, down to the Hellenistic era, into Judaism, and westwards to Hellenistic and other learned elites in the rest of the Mediterranean world.[101]

The cross-cultural and scientific connotations of the Aramaic language also—Ben-Dov suggests—serve to place the choice of Hebrew, instead, in 4Q186 and 4Q317 into sharper relief. If this choice reflects a

[98] Ben-Dov, *Head of All Years*, 6–7; idem, "Scientific Writings," 380-381.
[99] Ben-Dov, *Head of All Years*, 1.
[100] Ben-Dov, *Head of All Years*; idem, "Scientific Writings," 238-239.
[101] Ben-Dov, *Head of All Years*, 245-287, 140-146; idem, "Scientific Writings," 393-397.

renewed association of the Hebrew language with Jewish peoplehood and pedagogy, then it may simultaneously signal an emergent sense of "ancient Jewish sciences" as *Jewish*, perhaps concurrent with the evolving autonomy, systemization, and hyperrealism of the calendrical system; if the translation into Hebrew placed practical constraints on transmission, this limitation may have been deliberate, reflecting a self-conscious decision between [1] participating in an Aramaic discourse, at a nexus of translatability and transmission between multiple cultures, with a cosmopolitan horizon, and [2] creating a new Hebrew *qua* inner-Jewish discourse, marked by multiple levels of specificity and secrecy (e.g., Jewish peoplehood, priestly pedagogy, transmission among elite experts).

Popović remains skeptical about just how much can be inferred from 4Q186 and 4Q317.[102] His call for caution is well taken, particularly given the preliminary state of the present discussion. In the present context, however, we might hazard some speculations as to the ramifications if Ben-Dov is correct. Foremost for the history of Jewish knowledge is the possibility that the Second Temple period might have seen something akin to what Langermann identifies at the tail end of Late Antiquity as emergent "Hebrew scientific literature"[103]—with the embrace of Hebrew as a potential technical language, in anonymous or pseudonymous works using traditional literary forms and genres, concurrent with attempts to begin to create some distinctively Jewish synthesis, which may be part of a broader cross-cultural surge of interest in the cosmos but remains ultimately irreducible to "borrowing."

For the history of science more broadly, Ben-Dov's findings may also hold promise. For many decades, specialists in that field have stressed the constructedness and contingency of "science,"[104] sometimes even to the degree that the category might seem to be meaningful primarily as a modern or European invention. Peter Dear, for instance, thus questions whether the "essentialist universalism" of older studies[105] has now given way to the myopia of

102 Popović, "The Emergence," 105-6.
103 Langermann, "Hebrew Scientific Literature."
104 E.g. Kuhn, *The Structure of Scientific Revolutions*.
105 E.g. Sarton, *History of Science*.

"hyperhistoricism."[106] The focus on the specific contexts and dynamics of local phenomena is laudable. What might have been lost in the process, however, is some sense of "science" as constituted by a premodern as well as modern history, both marked by cross-cultural connectivity:

> ... there ought to be some way of speaking coherently about those knowledge enterprises ... that had a career that spread across the Eurasian continent, where Ptolemaic planetary models show up in seventeenth-century Jaipur as well as ninth-century Cologne. In such cases, techniques spread through adoption, rather like the apparently organic growth and decline of language groups, even though, as in the case of historical linguistics, we know that the spread is effected by countless episodes of human social interaction.[107]

Above, I suggested that it is precisely the element of interconnectivity —and the self-conscious negotiation of universality and difference often surrounding it—that makes modern "science" a potentially interesting analogy for exploring the practice and perception of "ancient Jewish sciences" in their broader Jewish and non-Jewish contexts. Particularly in light of Ben-Dov's work, moreover, we might reverse the arrow of comparison as well, bringing the example of ancient Judaism to bear on the problem of situating "science" simultaneously in local and trans-local contexts.[108]

In a recent essay on global history and early modern science, for instance, Lissa Roberts extends Dear's insights, stressing the need for historians of science to "attend to and connect two seemingly different orientations: the specifically local character of individual encounters and the increasingly global networks that both afforded and attributed meaning to the conditions and outcomes of these local exchanges."

[106] Dear, "What is the History of Science?" 406.
[107] Ibid.
[108] This historiographical point is made more broadly for the early modern period—also in relation to Jews and Judaism—by David B. Ruderman, *Early Modern Jewry: A New Cultural History* (Princeton: Princeton University Press, 2010), 12, with an emphasis on "the dialectical relationship between local conditions and continental or even global patterns."

She points to a number of recent studies that have begun such work for the modern period (esp. 18th and 19th centuries), positing that:

> ... in place of a view of science as the West's gift to the world or histories that focus on western science primarily as a tool of imperialist domination, a dynamically balanced approach is emerging which seeks to highlight the productive role played by globally situated intercultural exchanges in the history of science and history more generally, while simultaneously recognising the asymmetrical character of the conditions that often attended such encounters.[109]

As we have seen, Ben-Dov's analysis of the Jewish astronomical traditions at Qumran achieves something similar for a neglected set of premodern sources for the study of ancient sciences. He posits "ancient Jewish sciences" as an integral part of the networks of scientific knowledge interlacing the Hellenistic world, serving perhaps even as one of the channels by which information circulated westwards. Yet he also allows for its status as a "self-contained construct," articulated and practiced in local language, idioms, and aims—best studied as part of a broader cross-cultural network but also, simultaneously, "from within" Judaism.

Nevertheless, even as Ben-Dov makes a powerful argument for bringing ancient Jewish sources to bear on the history of science, his account still leaves "ancient Jewish sciences" on the margins of the history of Judaism. To be sure, he sidesteps the contrast between "religion" and "science," stressing that the two coexist inextricably intermingled in "ancient Jewish sciences" as in their Babylonian predecessors.[110] Whereas Popović maps different possible motives for Jewish engagement with sciences,[111] however, Ben-Dov locates the "cosmological imperative" of ancient Judaism largely in an apocalyptic impulse. Furthermore, he traces this trajectory in terms distinct from sapiential and other streams of Second Temple Judaism, and he asserts

[109] Roberts, "Situating Science in Global History."
[110] Ben-Dov, *Head of All Years*, 276-278.
[111] Popović, "The Emergence," 98.

their articulation in increasingly esoteric terms precisely in their Hebrew-language expressions.

It may be tempting to dismiss the various instances of scientific engagement within Jewish texts as exceptions to a religious tradition marked by a lack of any interest of this sort. Indeed, as we have seen, much of the discussion of "ancient Jewish sciences"—both past and present—seems to assume as much. Much cited, for instance, are Ben Sira's warnings about speculation into hidden things (3:21), the Mishnah's limitation on the exposition of cosmogony and cosmology (*m. Ḥagigah* 2.1), and the Talmudic use of the former to expound the latter (*y. Ḥagigah* 2.1/77c; *b. Ḥagigah* 13a).[112] It might seem natural, thus, to assume that any "cosmological imperative" among Enochic or Qumranic scribes must have been limited in time and influence, standing at some remove from the Judaism of the Torah and its Rabbinic interpreters—or, in other words, that "ancient Jewish sciences" have very little to tell us about Judaism.[113] Moreover, as we have seen, assumptions of this sort also dovetail with broader tendencies common in some recent synthetic approaches to Jewish intellectual history,[114] whereby the complexity of the ancient literary evidence is resolved with appeal to a series of dichotomies (e.g., Enochic vs. Mosaic, apocalyptic vs. sapiential, mystical vs. mainstream, priestly vs. Rabbinic), often overlaid tidily upon one another.

In what follows, I would like to suggest that a focus on "ancient Jewish sciences" might help us to recover some of this complexity in new and interesting ways, particularly if we situate recent findings concerning the Dead Sea Scrolls within a broader scope of Jewish literature and history. New insights from the study of astrology and physiognomy in Second Temple Judaism may enrich our understanding of the Jewish literatures of Late Antiquity, just as the late antique evidence may allow us to "test" these insights across broader historical trajectories within Judaism. By drawing

[112] Less often noted but no less intriguing—particularly for our present purposes—is the Rabbinic paraphrase of ben Sira's positive statements about physicians, etc. (e.g., Ben Sira 38:1ff; *y. Taʿanit* 3:6/66d; *Exodus Rabbah* 21.7).
[113] Cf. Neusner, "Why No Science in Judaism?"
[114] E.g., Boccaccini, *Beyond the Essene Hypothesis;* Elior, *The Three Temples*; Alexander, "Jewish Priesthood."

methodological insights from the work of Popović, in particular, we may be able to highlight some of the lines of continuity connecting Second Temple, Rabbinic, and early medieval Jewish writings about the cosmos.

3. *Maaseh bereshit, "science," and secrecy*

At first sight, the evidence of the classical Rabbinic literature might seem to resist any connection to "ancient Jewish sciences." In the case of Second Temple Judaism, we find texts in Hebrew, Aramaic, and Greek that intersect with non-Jewish discussions and debates about the origins and spread of cross-cultural forms of knowledge and expertise. Such concerns, as we have seen, find intriguing parallels in early medieval Jewish reflections on the place of "science" in Judaism, as attested in Hebrew and Judeo-Arabic literature from *Sefer Asaf* to Judah Halevi and beyond. The surviving Jewish literature of the period in between, however, is almost wholly Rabbinic,[115] and it is characterized by some notable breaks from the taxonomies and genealogies of knowledge that connect Hellenistic, Roman, and Islamic imperial discourses, and ancient and medieval Jewish engagement with them. There are notable literary and discursive differences both [1] from Second Temple Jewish discussions about the stars, cosmos, human body, and the history of knowledge, and [2] from the Mesopotamian, Hellenistic, and Roman traditions that such earlier discussions engaged.

As we have seen, the literary representation of ancient sciences in Second Temple Judaism can be read against the background of Hellenistic and Roman imperial claims about the history of knowledge, as well as in relation to Babylonian, Egyptian, and other counter-claims. Furthermore, just as a range of Jews in the Second Temple period seem to have engaged with types of knowledge that were cross-culturally cultivated and discussed, so Jewish intellectuals in the early Middle Ages variously engaged in scientific inquiries, and in debates about the prehistory of such inquiries, together with and

[115] *Piyyutim*, notably, present potentially important but underutilized sources for this broader discussion; see, e.g., Michael Rand, "Clouds, Rain, and the Upper Waters: From Bereshit Rabbah to the Piyyuṭim of Eleazar be-rabbi Qillir," *Aleph* 9 (2009): 13-39.

parallel to their Muslim counterparts and contemporaries. By contrast, discussions of the cosmos, stars, and human body in the Rabbinic literature of Late Antiquity largely defy interpretation in such terms, not least because of the Rabbinic innovation of new categories, which organize knowledge in ways that depart from the Second Temple Jewish precedents noted above and—perhaps not coincidentally—resist translation into the cross-cultural terms and non-Jewish taxonomies of their own time as well.

Most significant, for our purposes, is the category of *maaseh bereshit*, wherein the innovative character of the Rabbinic reconfiguration of Jewish knowledge is both achieved and effaced by a neologism that embodies claimed continuity with the Torah.[116] To my knowledge, there is no clear non-Jewish precedent or pre-Rabbinic counterpart for this category, which encompasses the divine creation of the cosmos (e.g., *t. Sanhedrin* 8.7–9), the account thereof in Genesis 1 (e.g., *m. Taanit* 4.2–3; *m. Megillah* 3.6), and the cosmic order resultant from it. [117] The term itself explicitly evokes Genesis, using the first word of that work (i.e., *bereshit*, "in-the-beginning") to telegraph creation and its products.

Yet if, at first sight, the Rabbinic development of such a Torah-based category even to describe and delineate knowledge about the cosmos might seem to offer a "parade example" of the subordination of "science" to "religion," closer examination reveals a more complex situation, which resists any easy reduction to modern

[116] Cf. Howard Eilberg-Schwartz, "Creation and Classification in Judaism: From Priestly to Rabbinic Conceptions," *History of Religions* 26.4 (1987): 357-381. The retrospectively normative status of the Mishnah and Talmud have meant that Rabbinic claims vis-à-vis continuity with the Hebrew Bible are often taken at face value. Helpful for our purposes are Martin Jaffee's insights into the reorganization of knowledge attendant upon the early Rabbinic articulation, defense, and naturalization of new modes of pedagogy in *Torah in the Mouth: Writing and Oral Tradition in Palestinian Judaism, 200 BCE-400 CE* (Oxford: Oxford University Press, 2001), 87–92. Through a focus on Rabbinic "curriculum pericopes" (e.g., *Mekhilta de R. Ishmael ad* Exodus 15:26; *Sifra Shemini* par.1:9 ad Leviticus 10:10–11), for instance, he highlights "a hermeneutical procedure in which scriptural terms are systematically re-signified and reconfigured so as to anticipate and define the rabbinic taxonomy of traditional learning" (p. 88).

[117] E.g., *m. Berakhot* 9.2; see further David J. Halperin, *The Merkabah in Rabbinic Literature* (New Haven: American Oriental Society, 1980), 19–63.

categories. When we first encounter *maaseh bereshit* in the Mishnah, it is in seemingly contrasting contexts. The term occurs in a number of mishnaic discussions of, and allusions to, the place of Genesis 1 and the visible world in Jewish liturgy and ritual practice (e.g., *m. Taanit* 4.2–3; *m. Megillah* 3.6), including calls to bless God as "the One who makes *maaseh bereshit*" when one sees geographical phenomena (mountains, hills, rivers, etc.; *m. Berakhot* 9.2). Yet the Mishnah and Tosefta also include strictures and stories about the dangers of expounding *maaseh bereshit*, in association with discussions of the dangers of speculation into what lies before and beyond the visible world.[118] It is the latter, as Alon Goshen-Gottstein has noted, that have attracted the most scholarly attention, and as a result, the topic of *maaseh bereshit* has been discussed almost solely in relation to Jewish mysticism.[119]

At the center of the discussion has been the famous mishnah in which *maaseh bereshit* appears alongside the *merkavah* (i.e., the divine chariot and description thereof in Ezekiel), in a curtailing of exegetical inquiry followed by the denunciation of prideful speculation into pre-creation, cosmology, and eschatology:

> It is not permitted to expound [*doreshin*] *aravot* among/to three, nor *maaseh bereshit* among/to two, nor the *merkavah* among/to one, unless he is wise and understands on his own.
>
> Anyone who speculates about four things, it would be merciful for him if he had not come into the world: what is above, what is below, what is before, what is after. Anyone who has no concern for the honor of his Creator, it would be merciful for him if he had not come into the world. (*m. Ḥagigah* 2.1)

Not only have scholars largely taken for granted that *maaseh bereshit* forms part of Jewish mysticism, but *m. Ḥagigah* 2.1 has been treated by some scholars as a key that unlocks the secret history of Jewish

[118] Esp. *m. Ḥagigah* 2.1; *t. Ḥagigah* 2.1–7. Cf. *y. Ḥagigah* 2.1/77a–c; *b. Ḥagigah* 11b–13a, 15a; Halperin, *The Merkabah in Rabbinic Literature*, 19–63; Schäfer, *The Origins*, 180-185, 207-210, 233-234.

[119] Alon Goshen-Gottstein, "Is *Ma'aseh Bereshit* Part of Ancient Jewish Mysticism?" *Journal of Jewish Thought and Philosophy* 4 (1995): 185–201.

mysticism as an esoteric and/or priestly movement evolving from the early Enochic materials discussed above into later Hekhalot traditions.[120] Just as the embrace of "mysticism" and "science" are sometimes paired in arguments for distinct Enochic and/or priestly stream of Judaism in Second Temple times, so resistance to both has been posited as characteristic of the Rabbinic tradition that is construed as its polar opposite, with ambivalence or antipathy towards visionary experience, etc., sometimes extended to "science" and cosmology, as exemplary of knowledge pursued apart from the Torah.[121]

Yet it is only with some difficulty that one constructs a monolithic "Rabbinic" opinion towards such issues, let alone an

[120] E.g. Nicolas Séd, *La mystique cosmologique juive* (Paris: Editions de l'Ecole des hautes études en sciences socials, 1981); Elior, *The Three Temples*, 211, for instance, reads the different elements in *m. Ḥagigah* 2.1 as clues to the specific characteristics of a single monolithic "mystical" movement opposed by the Rabbis, speculating that "the unexplained prohibitions imposed by the Sages in tractate *Ḥagigah* ... are precisely mirrored, in a striking fashion, by certain major obligations in the literature of the secessionist priesthood: the *Merkavah* as the cosmic prototype of the celestial Temple. .. ; *maaseh bereshit* as representing the totality of cosmological phenomena linking the sanctity of time and the solar calendar (four seasons, twelve months, seven days of the week, twenty-four hours of the day) with the sanctity of place and cult in a seven-based sequence guaranteeing the cycle of life with its correlated four fold and twelvefold divisions; and *ʿarayot*, sexual union, representing the body of traditions relating to holy union, the Temple, and holy matrimony ..." This characterization is certainly conceptually appealing, but it is not supported by the sources; see further, e.g., Halperin, *The Merkabah* (esp. 23) for a discussion of the meanings of *maaseh bereshit*, as based on a broader range of traditions from the classical Rabbinic literature. See also Alon Goshen-Gottstein, "*Maʿaseh Bereshit,*" on the ways that our understanding of the Rabbinic idea of *maaseh bereshit* can be skewed when it is examined only in the context of discussions of the merkavah and Jewish mysticism.

[121] Notably, Philip S. Alexander offers a more nuanced assessment of the Rabbinic evidence; see, e.g., Alexander, "Pre-emptive Exegesis: Genesis Rabba's Reading of the Story of Creation," *JJS* 43 (1992): 230-245; idem, "Enoch and the Beginnings," 229-230; cf. Neusner, "Why No Science in Judaism?"; idem, "Science and Magic, Miracle and Magic, in Formative Judaism: The System and the Difference," in *Religion, Science, and Magic In Concert and In Conflict* (ed. J. Neusner, E. S. Frerichs, and P. V. M. Flesher; New York: Oxford University Press, 1989), 61-81. Nevertheless, Alexander too ("Enoch and the Beginnings," 226) points to the strictures on *maaseh bereshit* as among the factors that "have inhibited serious Rabbinic involvement in science."

attitude or mindset consistently opposed to "science."[122] Just as the limits placed on human inquiry in biblical books such as Deuteronomy and Job are broached in apocalyptic and parabiblical works in the Enochic literary tradition, as well as in scientific texts from Qumran, so the constraints on cosmogonic and cosmological inquiry in the Mishnah are broached already within the Rabbinic literary tradition, in works like *Genesis Rabbah*[123] and the Talmud Bavli;[124] indeed, the Bavli's own concerns with cosmology provide the immediate context in which we must try to understand later Hebrew treatises like *Sefer Yetzirah*, *Midrash Konen*, and *Seder Rabbah di-Bereshit*. When one does find Rabbinic resistance to cosmological speculation, moreover, it is most often expressed—as Peter Schäfer has shown—through the appropriation of the ideas in question.[125] Rather than a rejection of "science," what we see is a more of a totalizing impulse to encompass all varieties of knowledge, while maintaining the epistemological monopoly of the Torah. If Rabbinic discussion about the cosmos and human knowledge about it resists the totalizing claims of the imperial taxonomies and genealogies of knowledge discussed above, it is by offering alternatives that are no less global in scope, remapping the cosmos in the image of the Torah and the Torah in the image of the cosmos.

Elsewhere, I have thus argued against the tendency to read the esotericism of the Rabbinic discussion of *maaseh bereshit* primarily in relation to a reaction against "mysticism," and I have proposed that it might be better understood in terms of the rhetoric of secrecy associated with calendrical and other specialist knowledge elsewhere

[122] Cf. Neusner, "Why No Science in Judaism?"
[123] Esp. *Genesis Rabbah* 1-8; cf. Alexander, "Pre-emptive Exegesis."
[124] Esp. *b. Ḥagigah* 12b–13a; cf. Peter Schäfer, "Bereshit bara Elohim: Bereshit Rabba, Parasha 1, Reconsidered," in *Empsychoi Logoi: Religious Innovations in Antiquity: Studies in Honour of Pieter Willem van der Horst* (ed. A. Houtman, A. de Jong, and M. Misset-van de Weg; Leiden: Brill, 2008), 267-289.
[125] Peter Schäfer, "In Heaven as It Is in Hell: The Cosmology of *Seder Rabbah de-Bereshit*," in *Heavenly Realms and Earthly Realities in Late Antique Religions* (ed. Ra'anan S. Boustan and Annette Yoshiko Reed; Cambridge: Cambridge University Press, 2004), 233-274; idem, "From Cosmology to Theology: The Rabbinic Appropriation of Apocalyptic Cosmology," in *Creation and Re-Creation in Jewish Thought: Festschrift in Honor of Joseph Dan on the Occasion of His Seventieth Birthday* (ed. Rachel Elior and Peter Schäfer; Tübingen: Mohr Siebeck, 2005), 39-58; idem, "Bereshit Bara Elohim".

in the Rabbinic literature.[126] I suggested that the rhetoric of secrecy surrounding *maaseh bereshit* might be likened to the discourse surrounding *sod ha-ibbur* (i.e, "the secret of the calendar"; *b. Rosh Hashanah* 20b), reflecting "an effort to articulate, isolate, and elevate a certain domain of expertise, while enhancing the intellectual prestige and 'religious' authority of those few who can master it."[127] If more recent research has revealed the richness of the scientific heritage from Second Temple Judaism that Rabbis might have inherited, it also raises the possibility that the secrecy surrounding *maaseh bereshit* and *sod ha-ibbur* might be something more than rhetoric. On the basis of the Qumran evidence, for instance, Ben-Dov suggests that "ancient Jewish sciences" depart from the public availability and circulation of knowledge often associated with modern science:

> ... the science of the Dead Sea Scrolls adheres to a different model, by which speculative wisdom is an esoteric venture, to be concealed from laymen and revealed to the initiated only. This esoteric science was the norm in ancient Mesopotamian literature, which ... was the source for a great part of what later appears in the scrolls.[128]

If Ben-Dov is correct, we might further wonder whether the Rabbinic discussion of *maaseh bereshit*, *sod ha-ibbur*, etc., might reflect a long-standing structural feature of the ways in which calendrical, astronomical, and/or other scientific knowledge was preserved and taught within Jewish cultures—whether in continuity with the emergence of Hebrew scientific literature at Qumran[129] or due to later points of contact with Babylonian sciences and pedagogy.[130] This is particularly the case if we follow the lead of Popović, accepting that

[126] Reed, "Was there Science in Ancient Judaism?" 476-482.

[127] Reed, "Was there Science in Ancient Judaism?" 479. For an example pertaining to medicine, see *b. Berakhot* 10b on the allegedly hidden *Sefer Refuot* of King Hezekiah.

[128] Ben-Dov, "Scientific Writings," 381.

[129] Samuel Thomas, *The "Mysteries" of Qumran: Mystery, Secrecy, and Esotericism in the Dead Sea Scrolls* (SBLEJL 26; Atlanta: SBL, 2009).

[130] Geller, "The Last Wedge."; idem, "The survival of Babylonian Wissenschaft;" idem, "An Akkadian Vademecum;" idem, "Akkadian Healing Therapies."

"secrecy does not only have to refer to a specific content of a body of knowledge or its comprehensibility" but may be "better understood as a means to organise the accessibility and availability of information and learning, and this in connection with the social status that it bestows on those possessing it."[131]

In support of such possibilities, we might adduce the scattered references in the classical Rabbinic literature to Sages said to be experts in astronomical, calendrical, and medicinal matters. Despite a lack of concerted or focused discussions of such concerns, for instance, one does find depictions of R. Gamaliel with lunar diagrams, showing witnesses to the phases of the moon in *m. Sanhedrin* 2.8. Six people are granted the title "physician" (i.e., *rofe* or *asya*), namely, Theodos/Theodoros, Tobiya, Bar Ginte, Minyomi/Benjamin, R. Ammi, and Bar Nathan.[132] Mar Samuel, in addition, is lauded as an expert in eye diseases (*b. Avoda Zara* 28b; *b. Shabbat* 78a, 108b; cf. *b. Bava Metsia* 85b), in the treatment of medical complications such as those arising from circumcision (*b. Shabbat* 133b–134b, 137a–b; *b. Ketubot* 110b; *b. Nedarim* 37b, 41a, 54b; *b. Gittin* 70a; *b. Bava Batra* 146a; *b. ʿAvodah Zarah*

[131] Popović, "Physiognomic Knowledge," 169-170; cf. Pamela O. Long, *Openness, Secrecy, Authorship: Technical Arts and the Culture of Knowledge from Antiquity to the Renaissance* (Baltimore: Johns Hopkins University Press, 2001).

[132] Of these, most is said about Theodos (*m. Bekhorot* 4.4; *t. Ahilot* 4.2; *b. Nazir* 52a; *b. Sanhedrin* 33a; *b. Bekhorot* 28b), whom some scholars have tried to connect, without much success, with references to figures named Theudas in the writings of Galen (e.g., *De Meth. Med.* 2.7 [10.142]; Andromachos in *Comp. med. Genera* 6.14 [13.925-926]; see further Julius Preuss, *Biblical and Talmudic Medicine* (trans. F. Rosner; New York: Sanhedrin Press, 1978 [1911]), 19–20; Samuel S. Kottek, "Alexandrian Medicine in the Talmudic Corpus," *Koroth* 12 (1996–1997): 85–87. Galen does, however, make use of the work of another Jewish physician, Rufus of Samaria (fl. ca. 100 CE), who lived in Rome and was well-known for Greek commentaries on the sixth book of Hippocrates' *Epidemics*; see further Galen in *Corpus Medicorum Graecorum* 5.10.2.2, pp. 213, 289, 293, 413; Franz Pfaff, "Rufus aus Samaria," *Hermes* 67 (1932): 356-359; S. Muntner, "Rufus of Samaria," *Israel Medical Journal* 17 (1958): 273-275; Annette Yoshiko Reed, "Asaph ha-Rofe," "Domnus," "Gamaliel VI," "Rufus of Samaria," "Samuel of Nehardea," "Theodos of Alexandria," and "Zakhalias of Babylon," in *Encyclopedia of Ancient Natural Scientists: The Greek Tradition and its Many Heirs* (ed. Paul T. Keyser and Georgia Irby-Massie; London: Routledge, 2008), 168, 275, 342-3, 721, 726, 788-789, and 843. For further Rabbinic references to physicians, therapies, etc., see also Geller, "Akkadian Healing," 1-60; Samuel S. Kottek, "Medical Interest in Ancient Rabbinic Literature," in *The Literature of the Sages, Second Part* (ed. Shmuel Safrai, et al; Minneapolis: Fortress, 2006), 485-496.

28a), in astronomy and bloodletting (*b. Berakhot* 58b; cf. *b. Shabbat* 129a–b, 156b), as well as in the information about lunar, solar, and planetary cycles needed for the calculation of the lunisolar calendar and intercalation of months (*b. Eruvin* 56a; *b. Rosh Hashana* 20b; *b. Sanhedrin* 12b; *b. Arakhin* 9b). In addition, even the seemingly routine exaltation of God as Creator is sometimes expressed with reference to surprisingly detailed matters of astronomical and calendrical cycles. In *b. Berakhot* 59a, for instance, one finds the injunction to bless God as "the One who makes *maaseh bereshit*" whenever one "sees the sun at its turning-point [*tequfah*], the moon in its power, the stars in their orbits, and the *mazzalot* in their orderly progress"—something that happens, according to Abaye, "every twenty-eight years when the cycle [*tequfah*] begins again and the *tequfat Nisan* (i.e., spring equinox) falls in Saturn on Tuesday evening, going into Wednesday" (cf. *Leviticus Rabbah* 23.8).[133] That the cultivation of expertise in such areas went beyond the specific Sages associated by name with astronomical or medicinal expertise, moreover, is suggested by the scattered but significant references to astrological tropes and medicinal therapies.[134]

In addition, some Rabbinic awareness of non-Jewish cosmology seems to be signaled by the *baraita* about planetary motion in *b. Pesahim* 94b, which culminates in an intriguingly positive assessment of non-Jewish knowledge about the cycles of the sun:

> Our Rabbis taught: The Sages of Israel maintain: "The *galgal* [celestial sphere?] is stationary, while the *mazzalot*

[133] The cosmogonic ramifications of this cosmological calculation is noted by Sacha Stern, "Fictitious Calendars: Early Rabbinic Notions of Time, Astronomy, and Reality," *JQR* 87 (1996): 105: "This 28-year cycle begins, according to bBer 59b, with the spring equinox (*tequfat Nisan*) occurring exactly at the beginning of the fourth day of the week (i.e., Tuesday evening). According to later sources, and as already implicit, perhaps, in the Babylonian Talmud, the first 28-year cycle began at the time of the world's creation; indeed, the spring equinox occurred in 'year 1' at the beginning of the same fourth day of the week as when, according to Gen 1:14–19, the sun was created (in other words, the sun was created in a position of spring equinox). The occurrence of any subsequent *tequfah* can thus be worked out with reference to the time of the sun's creation."

[134] Geller, "An Akkadian Vademecum"; Leicht, *Astrologumena Judaica*, 39–106; Kottek, "Medical Interest"; Mira Balberg, "Rabbinic Authority, Medical Rhetoric, and Body Hermeneutics in Mishnah Nega'im," *AJS Review* 35.2 (2011): 323-346.

> [planets?] revolve." But the sages of the nations of the world maintain: "The *galgal* revolves and the *mazzalot* are stationary" ... The Sages of Israel maintain: "The sun travels beneath the sky by day and above the sky at night." But the sages of the nations of the world maintain: "It travels beneath the sky by day and below the earth at night." Rabbi said: "And their view is preferable to ours, since wells are cold by day but warm at night."

Likewise, just as halakhic discussions about the human body sometimes recall physiognomic concepts and debates known from Greek and Roman scientific literature,[135] so some allusions to embryology in Rabbinic aggadot resonate with their Roman as well as Persian counterparts.[136] Some interaction of Palestinian sages with Roman healing practices may be signaled by the importation into Hebrew of Greek technical terms for some physician's tools,[137] and Talmudic allusions to therapies raise the possibility that some Babylonian sages may have been familiar with Babylonian healing practices otherwise known to us only from Akkadian sources.[138] Further examples of the sort might be cited in relation to other areas like botany, zoology, and geography.[139] What is important for our purposes, however, is that such references seem to attest some scientific interests, expertise, and information among late antique

[135] See further Charlotte Fonrobert, "The Semiotics of the Sexed Body in Early Halakhic Discourse," in *How Should Rabbinic Literature Be Read in the Modern World?* (ed. Matthew Kraus; Piscataway, NJ: Gorgias Press, 2006), 102.

[136] Abraham Ofir Shemesh, "Therapeutic Bathing in Rabbinic Literature: Halachic Issues and their Background in History and Realia," *Jewish Medical Ethics* 7.2 (2010): 510-511; Reuven Kiperwasser, "Three Partners in a Person: The Genesis and Development of Embryological Theory in Biblical and Rabbinic Judaism," *lectio difficilior* 2 (2009): 21–27 [http://www.lectio.unibe.ch].

[137] Kottek, "Medical Interest," 489.

[138] Geller, "Akkadian Healing Therapies."

[139] E.g., Abraham Ofir Shemesh, "Biology in Rabbinic Literature: Fact and Folklore." in *The Literature of the Sages, Second Part*, 509-519; Zeev Safrai, "Geography and Cosmography in Talmudic Literature," in *The Literature of the Sages, Second Part*, 497-508.

Sages, even despite the lack of sustained engagement with such topics within the classical Rabbinic literature.[140]

The classical Rabbinic literature provides little information for the student who might wish to follow in the footsteps of R. Gamaliel in lunar expertise or Mar Samuel in the treatment of disease, and it even provides few details on how precisely one learns to assess the relative value of different ideas about celestial cycles or to determine the equinox. One could infer, for instance, that the challenges of maintaining the lunisolar calendar through intercalation must have necessitated the cultivation of some pedagogical methods for preserving, teaching, and transmitting more information about calendrical astronomy than we now find recorded in the extant literary records of Rabbis from Late Antiquity.[141] That medieval authors must try to reconstruct the calculations behind Talmudic statements, however, only serves to emphasize the apparent separation of such domains of knowledge from other areas of Rabbinic

[140] Notably, studies of Rabbinic treatments of the full range of relevant topics—astronomy, astrology, medicine, biology, geography, mathematic, etc.—have made note of this pattern. Mark Geller ("Akkadian Healing Therapies," 4), for instance, cautions that "[a]ny references to medicine in the Talmud are purely coincidental and serendipitous, cited as aspects of daily life which were loosely associated with points of Jewish law or custom . . . . We never have a full medical text in the Talmud, but only fragments of such texts, often within an anecdotal context" (so too Kottek, "Medical Interest," 485). Shemesh ("Biology in Rabbinic Literature," 508) stresses with reference to biology, zoology, etc., that "[w]hile non-Jewish authors wrote books specifically on nature-related topics, the sages expressed their opinions on these topics in the framework of their religious-halakhic discussions. Consequently, reference to animals in the mishnaic and talmudic literature is random." Indeed, as a result, even those who wish to consider Rabbinic perspectives on such topics must first engage in anthological endeavors (as already Preuss in his 1911 *Talmudic Medicine*). Likewise, those who wish to analyze Rabbinic perspectives on the workings of the body, stars, etc., solely on their own terms must nonetheless draw upon non-Jewish and/or post-Talmudic traditions to make sense of what are often extremely terse statements; for an interesting articulation of this challenge, see now on Rabbinic embryology Gwynn Kessler, *Conceiving Israel: The Fetus in Rabbinic Narratives* (Philadelphia: University of Pennsylvania Press, 2009).

[141] Cf. Otto Neugebauer, "The Astronomy of Maimonides and its Sources," *HUCA* 22 (1949): 322-324; Eliyahu Beller, "Ancient Jewish Mathematical Astronomy," *Archive for History of Exact Sciences* 38 (1988): 51-66.

expertise and practice which came to be more richly preserved in writing, such as halakha.[142]

Was the astronomy, medicine, etc., of late antique Sages too *ad hoc*, local, or eclectic to warrant systematic preservation or transmission in written forms? Was it mostly a matter of the engagement of some Jewish intellectuals in the learning of their broader non-Jewish cultural contexts? Or should we imagine some more cohesive tradition(s) transmitted primarily through oral channels, perhaps only among select sets of Rabbinic or other experts, who fostered some sense of secrecy? Might such transmission been facilitated by didactic texts of the sort only known from the Second Temple period due only to the discovery of the Dead Sea Scrolls? To what degree did the authors of the later Hebrew scientific literature collect and preserve earlier oral and/or written traditions, and to what degree did they retroject their own scientific interests back into the tannaitic past? Might any lines of continuity stretch back even into Second Temple times?[143]

Even if such questions cannot be answered with any certainty, they do point to one of the most helpful contributions of recent research on "ancient Jewish sciences": they have profitably reoriented the discussion of science and ancient Judaism to focus not only on questions about "progress" (i.e., who discovered what first and before

[142] That halakha can be understood as "scientific" in the sense of an impulse to organize knowledge about the world in an orderly fashion, involving experience-based inferences and logic-based arguments, etc., has been stressed by Menachem Fisch, *Rational Rabbis: Science and Talmudic Culture* (Bloomington: Indiana University Press, 1997); Neusner, "Science and Magic," and others. It also underlies early efforts—perhaps ripe now for revisiting—to compile Rabbinic traditions about the workings of the human body in the Talmud (e.g., Preuss, *Talmudic Medicine*), as Lawrence Schiffman aptly reminded us in his response to an earlier form of this essay at the ISAW conference on which this volume is based.

[143] To be sure, any speculation about choices of oral and textual transmission runs up against the challenges posed by the complexity of Rabbinic textuality more broadly—on which see now Talya Fishman, *Becoming the People of the Talmud: Oral Torah as Written Tradition in Medieval Jewish Cultures* (Philadelphia: University of Pennsylvania Press, 2011). My point here is only that one's sense of the relative plausibility of one or another channel of transmission makes a big difference in how one imagines the state of scientific knowledge and engagement among late antique Rabbis.

whom), but also and particularly on questions of pedagogy (i.e., how, where, why, and by whom accumulated scientific knowledge was systematized, preserved, taught, and transmitted). Popović, in particular, has pressed for attention to choices of literary form and framing, as possible clues as to the different settings of "scientific" training and transmission;[144] he thus pushes us, not just to consider the content of the extant records of "ancient Jewish sciences," but to ask what their literary context might reveal about the "context of transmission of scholarly knowledge"—"what textual formats or genres of scientific writings are attested? And what sort of authorial strategies did ancient Jewish scholars pursue?"[145]

Popović thus pinpoints critical questions, not just for the early materials, but maybe for later ones too. It may be tempting to search our literature for clues to the origins of scientific "attitudes," "interest," or "thought" within Judaism, but it is unclear whether literary evidence can even answer such questions.[146] What it can help to address, however, is the question of when, where, and how such interests came to be expressed explicitly in written forms and integrated into Jewish scribal and literary cultures. The extant data, after all, offer us only a small window onto the actual content of the knowledge about the cosmos, stars, and human body that circulated among ancient Jews, but richly attest the range of Jewish perspectives on the proper scope of human knowledge, the purposes of Hebrew writing, and the ideal purview and aims of investigating the cosmos.

Approached from this perspective, what is striking are the very different ways in which even seemingly recurrent concerns found expression in written forms in Second Temple, Rabbinic, and early medieval Jewish literature. Post-Talmudic works like *Baraita de-Shmuel*, *Sefer Yetzirah*, *Mishnat ha-Midot*, *Yetzirat ha-Walad*, *Pirqe de Rabbi Eliezer*, *Midrash Konen*, and *Sefer Asaf ha-Rofe*, for instance, claim continuity with the biblical and tannaitic past to varying degrees. All of them

[144] Popović, "The Emergence," 83.
[145] See Popović in this volume.
[146] On the limitations of cognitive approaches to the history of science, and their special dangers when paired with potentially reified cultural identities ("Jewish," "Greek," "Chinese," "Western," etc.), see Francesca Rochberg, "A Consideration of Babylonian Astronomy within the Historiography of Science"; Brennan, "The Birth".

depart strikingly from the Hebrew Bible and classical Rabbinic literature, however, inasmuch as they discuss astronomy, cosmology, mathematics, and medicine in explicit and focused fashion. It is this feature that led Langermann to draw attention to such works as neglected evidence for the "first chapter in the history of Hebrew scientific literature ... sometime during the very late Umayyad and early Abbassid caliphates," even as he stressed that "I am *not* asking when Jews first began to evince an *interest* in science. The question I want to answer is when Jews first *wrote* Hebrew texts whose primary purpose was the exposition of scientific knowledge."[147]

Langermann thus makes a critical distinction akin to that raised by Popović with respect to the relevant Qumran materials. In the Dead Sea Scrolls, as we have seen, we find evidence for emergence of Jewish cosmography first in relation to apocalyptic literary production in Aramaic but also perhaps in Aramaic and Hebrew didactic texts dedicated wholly to topics like physiognomy and astrology. In the case of the post-Talmudic works in which scientific interests again find sustained written expression, it is first with a spate of anonymous or pseudonymous writings in Hebrew, primarily framed in anthological genres modeled on the Mishnah, prior to and perhaps preparatory for the later emergence of authored treatises in Hebrew more readily recognized—by medieval Islamic as well as modern Western criteria—as "science."[148]

Are there any lines of continuity between "ancient Jewish sciences" and later Jewish literatures? And is there anything that we might learn about the seemingly sudden rise of a focused concern for the stars, human body, and structures of the cosmos in Aramaic and Hebrew literary cultures, in the wake of Hellenistic conquests in the Second Temple period, by looking also to the seemingly sudden rise of the same concerns within Hebrew literary cultures, many centuries later, in the wake of Islamic conquests?

As noted above, many of the relevant later works claim to contain (or were received as containing) faithful records in writing of the knowledge transmitted in secret by Sages of earlier times. *Seder*

[147] Langermann, "Hebrew Scientific Literature," 169-170; emphasis mine.
[148] Cf. Sela, *Abraham Ibn Ezra*, 7.

Rabbah di-Bereshit, for instance, circulated as the revelation of the secrets of *maaseh bereshit* mentioned by the Mishnah.[149] *Baraita de-Shmuel* is associated with the contents of Mar Samuel's astronomical wisdom, as possibly connected particularly to the *baraita* of the "secret of the calendar" [*sod ha-ibbur*] mentioned in the Talmud. In the hexaemeral chapters of *Pirqe de Rabbi Eliezer*.[150] Traditions about pre-creation, the substance from which earth and heaven were created, and the creation of heaven, angels, and the throne of God are attributed to R. Eliezer (*PRE* 3–4), while R. Yehoshua and R. Yehuda are associated with information about the ground, plants, earth, sun, moon, planets, zodiacal signs, and calendar (*PRE* 5–6), R. Eliezer and R. Meir with birds, fish, and insects (*PRE* 9). The most extensive astronomical materials, moreover, are here credited jointly to R. Yoḥanan ben Zakkai, R. Gamaliel, R. Ishmael, R. Eleazar ben Arach, R. Eliezer, and R. Akiva (*PRE* 7–8).

One would certainly not wish to read these claims at face value, suggesting—as did Nicholas Séd—that all these works reflect some single unified tradition of Jewish cosmology, attested in the apocalypses of Second Temple times, condemned by the Mishnah, and transmitted in secret until finding preservation in writings in texts like *Midrash Konen* and *Seder Rabbah di-Bereshit*.[151] Nor does Elior's theory of a "secessionist priesthood" that paired ritual concerns and mystical practice with the "mathematization of the cosmos" hold much explanatory power for works of this sort, which draw from the Torah and Talmud Bavli as much as (if not more than) from the Book of Ezekiel and the early Enochic literature.[152] Despite the allure of the quest to uncover a single esoteric movement running subterranean through ancient Jewish history, there is little to support the contention of a unified social or intellectual tradition connecting such

[149] Séd, *La mystique cosmologique juive*, 79–106; Schäfer, "In Heaven as It Is in Hell," 234. That *Seder Rabbah di-Bereshit* circulated primarily under the title *Ma'aseh Bereshit*, e.g., is clear from MS Munich 22. For the text see Peter Schäfer, ed., *Synopse zur Hekhalot-Literatur* (Tübingen: Mohr Siebeck, 1981), §§429–462, §§518–524, §§743–784, §§832–854.

[150] See further Reed, "Samuel of Nehardea"; eadem, "From Pre-Emptive Exegesis."

[151] Séd, *La mystique cosmologique juive*.

[152] Elior, *The Three Temples*.

materials, nor systematically distinguishing them from Rabbinic Judaism.[153]

Nevertheless, just as scientific materials from Qumran and beyond attest the surprisingly long and winding afterlives of elements from ancient Near Eastern astronomical and divinatory disciplines, so we must ask whether any threads of continuity might link the scientific interests attested in the Dead Sea Scrolls with those attested at the very end of Late Antiquity. The appearance of parallels between them, notably, forms part of the broader puzzle of the reappearance of Second Temple Jewish traditions in early medieval Jewish literature—some examples of which were noted above, with reference to *Sefer Asaf*'s parallels with Jubilees and Judah Halevi's possible knowledge of the *Wisdom of Solomon*.[154] In past research, the investigation of such connections was largely motivated by the search for evidence to support Gershom Scholem's passing musings about the possible prehistory of Jewish mysticism in the apocalyptic literature of Second Temple times.[155] If the discovery of the Dead Sea Scrolls initially seemed to hold out hope for providing evidence for Jewish mysticism's hidden heritage, however, many decades of concerted efforts to uncover concrete connections have achieved little more than the multiplication of impressionist parallels.

The question of possible connections between Enochic literature and later Jewish mysticism remains puzzling, but recent discussions of "ancient Jewish sciences" may help to provide a fresh perspective for further investigation. That a focus on physiognomy might help us to reorient the discussion in more useful ways, for instance, was suggested already by Michael Swartz in an essay from 2001 reviewing and reassessing the evidence for parallels between the Dead Sea Scrolls and Hekhalot literature. Swartz's survey and analysis distinguishes between different types of materials for which parallels have been noted in the past (what he calls "mystical," "magical," and "divinatory"), with an eye to the functions of the different types of

[153] Boustan, "Rabbinization."

[154] On this broader issue, see further references and discussion in Reed, *Fallen Angels*, 233-272.

[155] Gershom Scholem, *Major trends in Jewish Mysticism* (Jerusalem: Schocken, 1941).

knowledge and practice encompassed by each. What he shows is that those materials typically deemed "mystical" reveal less trans-historical connections than those that have been studied under the rubrics of "magic" and "divination." The latter, in fact, especially stands out: "the links between the physiognomic literature at Qumran and those from esoteric circles in late antiquity and the early middle ages are quite strong":[156]

> ... the magical and divinatory traditions have undergone the least change. It is also easier to identify the role that the latter play in the life of the community. This factor should give us pause to think about how we look at the history of Jewish spirituality in antiquity. While we have become accustomed to seeking visions—that is, looking for evidence for visionary practices at Qumran and in the Rabbinic milieu—we may have been ignoring another important source of revelation and divine disclosure. If the divinatory tradition is more pervasive and recognizable, we might reconsider our view of Qumran sectarians and Rabbinic esotericists, and perhaps of their contemporaries, as given to charismatic enthusiasms, and consider how they engaged in disciplined, intricate forms of reading not only of the *Sefer he-Hagiu*, but of the *Sefer Toledot Adam*.[157]

Swartz's findings prove significant, for our purposes, inasmuch as they point us both to the value of distinguishing different elements when theorizing the relationship between the Second Temple and later Jewish traditions, and to potential significance of "ancient Jewish sciences" for this task. Rather than continuing to debate the existence of some singular non-/anti-Rabbinic movement at the root of Jewish mysticism, it might prove more profitable to untangle the various threads that seemingly link the early Enochic literature and the Dead Sea Scrolls with later Jewish literary cultures, with an eye to the possible social settings and functions for the cultivation and

[156] Michael D. Swartz, "The Dead Sea Scrolls and Later Jewish Magic and Mysticism," *DSD* 8 (2001): 192.
[157] Swartz, "The Dead Sea Scrolls," 193.

transmission of different types of knowledge.[158] We know of a number of cases in which Second Temple Jewish traditions were seemingly lost or abandoned, but later recovered by Jews in the Middle Ages due to "back-borrowing" from Christian, Manichean, or Islamic tradents—as in the case of the circulation of the *Wisdom of Solomon* in Syriac among medieval Jews, or the recovery of the writings of Josephus; it is debatable whether something similar occurred in the case of other texts not preserved by late antique Rabbis, such as the *Book of Watchers* and Jubilees.[159] In some cases, certain motifs from these texts seem to have continued to circulate among Jews by virtue of their connection to various types of specialized knowledge that we might call "magic," and in other cases, due to their continued place in the oral-interpretative traditions surrounding certain biblical terms or verses.[160]

A focus on "ancient Jewish sciences" may help us further to illumine such dynamics. With the benefit now of Popović's work, it may be worth revisiting Swartz's insights into physiognomy as a practice possibly linking Second Temple Jewish reflection on the human body with its non-Jewish counterparts but also its later Jewish heirs. In the course of his comprehensive analyses of 4Q186 and 4Q561, he makes note of a number of terminological parallels with later materials, including passages in the Talmud Bavli, the post-Talmudic tractate *Baraita de-Mazzalot* and the *Physiognomy of R. Ishmael*, and various materials preserved in the Cairo Genizah.[161] The possibility that scientific traditions of this sort may have provided one possible channel for the transmission of some Second Temple traditions into later Judaism is also suggested by the intriguing evidence of a Genizah fragment preserving an apparently pre-Rabbanized form of *Sefer*

[158] I suggested something similar with respect to the transmission and circulation of angelological tropes via "magical" traditions in Reed, *Fallen Angels*, 253-255.
[159] See further Himmelfarb, "Some Echoes of Jubilees"; Reed, *Fallen Angels*.
[160] E.g., angelological motifs; Reed, *Fallen Angels*, 253-255.
[161] Popović, *Reading the Human Body*, 36–37, 105, 274-275, cf. 44 n. 105, 266 n. 2; also I. Gruenwald,. "New Fragments from the Physiognomic and Chiromantic Literature" [Hebrew], *Tarbiz* 40 (1970–1971): 301-319; Peter Schäfer, "Ein neues Fragment zur Metoposkopie und Chiromantik," in *Hekhalot-Studien* (Tübingen: Mohr Siebeck, 1988), 84-95.

Hekhalot, the Hekhalot text commonly called 3 Enoch.[162] Whereas a focus on Enoch has lead scholars since Hugo Odeberg to suggest that *Sefer Hekhalot*/3 Enoch stands in an unbroken line of development with Second Temple works like 1 Enoch and 2 Enoch, attention to the physiognomic concerns in its earliest attested form might help us to trace some of the separate threads that connect it with earlier tradition, apart from an expectation of simply proving or disproving direct filiation with earlier Enochic literature.[163]

Attending also to Popović's concerns for the literary framing and setting of knowledge, we might point to the enduring place of Genesis as a model for a number of the texts and traditions discussed above. The *Astronomical Book*, *Book of Watchers*, Jubilees, and Pseudo-Eupolemus, of course, represent some of our earliest evidence for the parabiblical tradition surrounding Genesis. If Josephus' discussions of Abraham as Chaldean scientist can similarly be read in this fashion, so perhaps too with the later Rabbinic discussion of astrology with reference to Abraham as well. It may be telling, moreover, that Genesis 5:1 is re-read as precedent and proof-text for Jewish physiognomy,[164] roughly around the same time that Enoch reappears as an emblem of cosmological knowledge in *Sefer Hekhalot*/3 *Enoch* and that Noah reappears as an emblem of medical knowledge in *Sefer Asaf ha-Rofe*.

Attention to this particular trend in the Jewish form and framing of scientific knowledge, moreover, highlights one of the most striking threads of continuity linking the relevant Second Temple, Rabbinic, and early medieval materials—namely, the use of the seven days of

[162] T.-S. K. 21.95L; Peter Schäfer, ed., *Geniza-Fragmente zur Hekhalot-Literatur* (TSAJ 6; Tübingen: Mohr Siebeck, 1984), 136-137; see also idem, *The Hidden and Manifest God: Some Major Themes in Early Jewish Mysticism* (Albany: SUNY Press, 1992), 147-148.

[163] As in the case of angelology, this may be a case of concerns that are attested also in earlier Rabbinic materials, in a less focused fashion, here developed in new directions. Discussing tannaitic traditions related to signs on the body vis-à-vis gender and its blurring, Charlotte Fonrobert ("Semiotics," 102), for instance, remarks on "what appears as a close relationship between the halakhic literature on semiotics of the body and late antique physiognomy," wherein parallels with Greek and Roman discussions occur despite a notable difference in orientation. See now Mira Balberg, "Rabbinic Authority."

[164] Popović, *Reading the Human Body*, 180, 277.

creation (Genesis 1:1–2:4) as a structure into which to integrate knowledge about the stars, cosmos, and human body gained *both* from biblical exegesis *and* from observation of the visible world and speculation into what lies beyond. One finds the expansion of the hexaemeron already with Jubilees, where it marks the "translation" of earlier Enochic materials in more ways than one—both from Aramaic into Hebrew, and also into the form of pentateuchal narrative. In one recension of *2 Enoch*, one finds a similar translation of sorts, with Enochic traditions now rendered in Hellenistic terms, but again with appeal to the hexaemeron; as in the *Book of Watchers*, Enoch here ascends to heaven, and in this case through multiple heavens, the contents of which are explained with appeal to cosmology, angelology, and eschatology. What is perhaps most striking, however, is what is revealed to him in the highest heaven—an account of God's creation of cosmos in seven days, retold in a manner that "updates" Genesis 1 but also remains within its bounds.

In Rabbinic midrash, such as *Genesis Rabbah*, the interpretation of Genesis 1 is shaped by new efforts to theorize of the bounds of proper inquiry into *maaseh bereshit*, in conversation with the Mishnah and Talmud Yerushalmi, and possibly in reaction to competing "scientific-cosmological" approaches.[165] That some association of the hexaemeron with scientific knowledge perhaps remains, however, might be suggested by an intriguing Talmudic reference to a passage from R. Joshua b. Levi's notebook, wherein the character and fate of a person were foretold based on the day of the week of his or her birth, in a manner intertextually connected to the cosmogony days described in Genesis 1 (*b. Shabbat* 156a).

Also notable, in this regard, is the calculation of *tequfot* associated with Abaye in *b. Berakhot* 59a, as quoted above. This calculation pertains to what is later explained more explicitly in later works like *Baraita de-Shmuel* (5–6) and *Pirqe de Rabbi Eliezer* (ch. 7) as *tequfat Shemuel*—one of the two main methods for calculating solistices and equinoxes, assuming a mean tropical year equal to that of the Julian calendar (365¼ days) and the beginning of the cycle on the day on

[165] Alexander, "Pre-emptive exegesis"; Schäfer, "Bereshit bara Elohim," 287-288.

which the sun was created according to Genesis (i.e., "Tuesday evening, going into Wednesday" = at the start of the fourth day). As such, it exemplifies the degree to which the Rabbinic discourse about *maaseh bereshit* cuts across modern categories of "science" and "religion"; the calculation can be deemed "scientific" in the sense of trying to make sense of observed patterns in nature by means of a theoretical model, but its meaning is not exhausted by this aim, and it would be misleading to judge its value solely in terms of modern notions of "science." Yet, as Sacha Stern has shown, one cannot understand such a system without a sense of the premodern creation of calendars as a practice for which accuracy was not the sole or even main criterion. With respect to *tequfat Shemuel*, in particular, Stern points to its conceptual value for conveying the orderliness of divine creation as well as its possible practical value as a mnemonic device.[166] To these, we might add its function as a midrash of sorts: consistent with the approach to the Torah as blueprint for creation in *Genesis Rabbah*, the cycles of the cosmos are here read through the lens of the first chapter of Genesis, thereby naturalizing the connection between all of the various senses of *maaseh bereshit* (i.e., as the divine process of creation and the biblical narrative about it, but also the visible products thereof).

Attention to issues of literary form and framing, however, also points us to the limits of midrash and *maaseh bereshit* to describe Rabbinic engagement with knowledge about the cosmos. Schäfer, for instance, has pointed to the dominance of another framework for organizing exegetical, observational, and speculative knowledge about the cosmos, namely, the seven heavens.[167] Despite seemingly belonging more to the world of ancient Jewish apocalypses than that of late antique Rabbis, moreover, this framework is determinative in the most detailed discussion of the cosmos in the Talmud Bavli (i.e., *b. Ḥagigah* 12b–13a). Partly as a result, moreover, the contents of the seven heavens are richly discussed across a surprisingly broad range of Rabbinic and para-Rabbinic materials—including *Pesiqta de-Rav Kahana*, *Leviticus Rabbah*, *Deuteronomy Rabbah*, and *Avot de-Rabbi Nathan* (A), but

[166] Stern, "Fictitious Calendars."
[167] Schäfer, "In Heaven as It Is in Hell;" idem, "From Cosmology to Theology."

also *Sefer Hekhalot/3 Enoch*, *Reuyot Yehezqel*, Targumic Tosefta to Ezekiel 1, *Midrash Konen*, and *Seder Rabbah di-Bereshit*.

Interestingly, it is not until the era that Langermann deems the "beginnings of Hebrew scientific literature" that the seven days return as a major organizational principle for Jewish knowledge about the natural world—whether together with the seven heavens or in place of it. In *Seder Rabbah di-Bereshit* and *Midrash Konen*, for instance, hexaemeral retellings function to frame and justify subsequent speculations about the contents of the seven heavens above, the seven earths below, classes of angels, and divine Throne. Perhaps more interesting, for our purposes, is the case of *Pirqe de Rabbi Eliezer*. There, the schema of the seven heavens (cf. PRE 18) seems to be set aside in favor of an expansive retelling of Genesis 1 (esp. PRE 3–11), which interweaves astronomical, meteorological, calendrical, geographical, and even zoological materials, alongside midrashic and aggadic traditions that recall Jubilees no less than *Genesis Rabbah*.[168] Not only does *Pirqe de Rabbi Eliezer* appeal both to primeval figures and Rabbis to justify its integration of "scientific" information (e.g., tracing intercalation to Adam, while associating R. Eliezer and his colleagues with teachings of astronomy, etc.), but its treatment of the fourth day includes substantial astronomical material, largely paralleling the system in *Baraita de Shmuel* and reflecting a written expression of calendrical astronomy more explicit than in earlier traditions.

In *Pirqe de Rabbi Eliezer*, moreover, ethical, ritual, and "scientific" materials are all presented in terms of a *Listenwissenschaft* that raises intriguing possibilities of some connection to pedagogical practice. Through numbered lists, the cycles and principles of Jewish piety are depicted as part of the divine order that permeates, enlivens, and supports the entire created world.[169] If cosmological inquiry is justified with appeal to the Torah, so Jewish ethics and practice are naturalized

[168] Annette Yoshiko Reed, "From 'Pre-Emptive Exegesis' to 'Pre-Emptive Speculation'? *Ma'aseh Bereshit* in *Bereshit Rabbah* and *Pirqei deRabbi Eliezer*," in *With Letters of Light-Otiyot Shel Or: Studies in Early Jewish Apocalypticism and Mysticism in Honour of Rachel Elior* (ed. D. Arbel and A. Orlov; Berlin: de Gruyter, 2010), 115-132.

[169] Annette Yoshiko Reed, "Who Can Recount the Mighty Acts of the Lord? Cosmology and Authority in Pirqei deRabbi Eliezer 1–3," *HUCA* 80 (2009): 115-141.

with appeal to the cosmos. If it is difficult to extricate "scientific" and "religious" aims here, it is also difficult to try to isolate the meaning of the form or framing of the knowledge itself from the meaning of the knowledge itself; the medium is part of the message.

4. Conclusion

Recent work on "ancient Jewish sciences" has much to tell us about the history of science, but it may speak to the history of Judaism, even beyond Second Temple times—not least by reminding us of what is lost when the history of Jewish knowledge is reduced to "religious" concerns or bifurcated along the lines of modern dichotomies. The data resist any easy reduction to familiar dichotomies like "religion" vs. "science," Near Eastern vs. Greek, or Jewish vs. foreign. Inasmuch as research on them necessarily crosses a number of conceptual and disciplinary divides in modern scholarship, moreover, it may also help open the way for more integrative perspectives on ancient Judaism.

We have seen, above, how traditions about the representation of the place of Jews in the history of knowledge serves as one thread connecting "ancient Jewish sciences" with their counterparts at the end of Late Antiquity. The connection is perhaps closest in the case of Jubilees and *Sefer Asaf ha-Rofe*, but even the less direct parallels share the appeal to the Hebrew Bible to explain the cross-cultural transmission of knowledge and the place of Judaism within it. The concern to do so, moreover, often seem shaped by the aim to argue that the history of Jewish knowledge encompassed the same topics as those of other cultures, albeit with a claim to extreme antiquity far greater than that of the Greeks, sometime paired with a claim to an unbroken tradition of written transmission rivaling even those of the Babylonians and Egyptians, and/or to a unique source for confirming the truth of such knowledge in angelic revelation, heavenly books, or eye-witness accounts of otherworldly realms. In both periods, new claims concerning the scope of Jewish scribal expertise may have also formed part of a defense of the ancient literary heritage of Israel—first against the challenges posed by the intellectual prestige of Hellenistic pedagogy and the cosmopolitan scientific discourse forged by learned elites in the Hellenistic Near East, and later against the challenges posed by the integration of Hellenistic sciences into Arabic pedagogy, translated and extended into a new cosmopolitan scientific discourse in the Islamic Near East.

Another point of possible continuity, as we have seen, is a model of scientific pedagogy and expertise, perhaps Babylonian in ultimate origin, as involving more secrecy than other forms of Jewish knowledge–a model that might remain active in practice, at times, but may also inform the continued appeal to the rhetoric of secrecy to add the luster of esotericism to claims of expertise and authority. If many of the materials surveyed above are marked by the claim to reveal what has long been secret, this claim may root its plausibility and power in historical moments in which older knowledge was being preserved systematically in written forms. Just as Enochic scribes preserve older Mesopotamian knowledge at the dawn of the early Hellenistic age, so too at the flowering of Islam–what some of the first Hebrew scientific writings (e.g., *Sefer Asaf*) preserve is not the new Islamic knowledge, but rather older sciences, including Hellenistic traditions.

Yet the cross-cultural and cosmopolitan horizon of much of the discussion—both in premodern texts and in modern scholarship—might lead us to overlook inner-Jewish concerns that may be no less significant for understanding "ancient Jewish sciences" and the various threads of their possible late antique and medieval afterlives. In the relevant later materials, as perhaps already at Qumran, the distinction of "science" and "religion" is more misleading than heuristic for understanding the ways in which different types of knowledge were organized, systematized, preserved, and transmitted. It may be telling, moreover, that the Rabbinic category of *maaseh bereshit* cuts directly across topics that might strike us—from a modern Western perspective—as obviously "religious" and topics that might strike us as obviously "scientific," encompassing exegesis of Genesis 1, observation of the visible world, and speculation about the creation and structures of the cosmos (cf. Maimonides, *Guide* I, 6–7; Judah Halevi, *Kitab al Khazari* IV 25).

If the Rabbinic discourse about *maaseh bereshit* thus cautions us against casually retrojecting our own assumptions about the mutual exclusivity of "religion" and "science" into earlier texts and periods, it also offers a sustained, self-conscious reflection on the bounds of proper Jewish knowledge and inquiry, and it reveals an important inner-Jewish dynamic—namely, the generative tension between the cosmographical reticence of the Torah itself, on the one hand, and the cosmological curiosity resultant from its claims that the God of Israel is the Creator of the cosmos, on the other. Whether or not one finds any pre-Rabbinic counterpart to the category of *maaseh bereshit*, it is worth wondering whether a similar dynamic is at play in Second Temple materials—perhaps somewhat neglected by virtue of the primary scholarly focus on whether and how "ancient Jewish sciences" relate to their non-Jewish counterparts and to modern standards of "science." If so, the ancient and late antique sources may offer overlooked materials for the history of science, but also understudied resources for the history of Judaism, valuable perhaps precisely because they might strike us as paradoxical in their visions of "ancient Jewish sciences."

A Bibliography for Ancient Jewish Sciences

Albani, Matthias 1994. *Astronomie und Schöpfungsglaube: Untersuchungen zum astronomischen Henochbuch* WMANT 68. Neukirchen-Vluyn, Neukirchener Verlag.

Albani, Matthias 1999. "Horoscopes in the Qumran Scrolls," in *The Dead Sea Scrolls after Fifty Years: A Comprehensive Assessment*. Eds. J.C. VanderKam & P.W. Flint, Leiden: Brill, 279-330.

Alexander, Philip S. 1992. "Pre-emptive Exegesis: Genesis Rabba's Reading of the Story of Creation," *Journal of Jewish Studies*, 43: 230-245.

Alexander, Philip S. 1996. "Physiognonomy, Initiation and Rank in the Qumran Community," in *Geschichte-Tradition-Reflexion: Festschrift für Martin Hengel zum 70. Geburtstag*. Ed. P. Schäfer, 385-394.

Alexander, Philip S. 2002. "Enoch and the Beginnings of Jewish Interest in Natural Science," in *The Wisdom Texts from Qumran and the Development of Sapiential Thought*. Ed. C. Hempel, A. Lange & H. Lichtenberger. BETL 159. Leuven: Peeters. 223-243.

Alexander, Philip S. 2009. "What Happened to the Jewish Priesthood After 70?" in *A Wandering Galilean: Essays in Honour of Seán Freyne*. Ed. Z. Rodgers, M. Daly-Denton, and A. Fitzpatrick McKinley. JSJSup 132. Leiden: Brill. 3-34.

Annus, Amar 2010. "On the Origin of Watchers: A Comparative Study of the Antediluvian Wisdom in Mesopotamian and Jewish Traditions," *Journal for the Study of the Pseudepigrapha*, 19: 277-320.

Argall, Randall A. 1995. *1 Enoch and Sirach: A Comparative and Conceptual Analysis of the Themes of Revelation, Creation, and Judgment*. Atlanta: Scholars Press.

Auffarth, Christoph & Loren T. Stuckenbruck (eds.) 2004. *The Fall of the Angels*. TBN 6. Leiden: Brill.

Bachmann, Veronika 2009. *Die Welt im Ausnahmezustand: eine Untersuchung zu Aussagegehalt und Theologie des Wächterbuches (1 Hen 1-36)*. BZAW 409. Berlin: de Gruyter.

Balberg, Mira 2011. "Rabbinic Authority, Medical Rhetoric, and Body Hermeneutics in Mishnah Nega'im," *AJS Review* 35: 323-346.

Basalla, George 1967. "The Spread of Western Science," *Science* 156: 616-622.

Baumgarten, Albert I. 2007. "Information Processing in Ancient Jewish Groups," in *Sectarianism in Early Judaism: Sociological Advances*. Ed. D.J. Chalcraft. London: Equinox, 246-255.

Baumgarten, Joseph M. 2003. "Qumranic and Astrological Terminology in Musar leMevin," *Tarbiz* 72: 321-328 (Hebrew).

Bedenbender, Andreas 2000. *Der Gott der Welt tritt auf dem Sinai. Enstehung, Entwicklung und Funktionsweise der frühjdischen Apokalyptik*. ANTZ 8. Berlin: Institut Kirche und Judentum.

Beller, Eliyahu 1988. "Ancient Jewish Mathematical Astronomy," *Archive for History of Exact Sciences* 38: 51-66.

Ben-Dov, Jonathan 2008. *Head of All Years: Astronomy and Calendars at Qumran in their Ancient Context*. STDJ 78. Leiden: Brill.

Ben-Dov, Jonathan 2009. "Hebrew and Aramaic Writing in the Pseudepigrapha and the Qumran Scrolls: The Ancient Near Eastern Background and the Quest for a Written Authority" (Hebrew), *Tarbiz* 78: 27-60.

Ben-Dov, Jonathan 2010. "Scientific Writings in Aramaic and Hebrew at Qumran: Translation and Concealment," in *Aramaica Qumranica: Proceedings of the Conference on the Aramaic Texts from Qumran in Aix-en-Provence 30 June - 2 July 2008*. Ed. K. Bertholet & D. Stökl Ben Ezra. STDJ 94. Leiden: Brill. 379-399.

Ben-Dov, Jonathan 2011. "The Qumran Dial: Artifact, Text, and Context," in *Qumran und die Archaeologie*. ed. J. Frey, C. Claussen and N. Kessler. WUNT 278. Tübingen: Mohr Siebeck, 211-237.

Ben-Dov, Jonathan 2012. "Lunar Calendars at Qumran? A Comparative and Ideological Study," in *Living the Lunar Calendar*. Eds. J. Ben-Dov, W. Horowitz & J. Steele. Oxford: Oxbow, 173-189.

Ben-Dov, Jonathan & Wayne Horowitz 2005. "The Babylonian Lunar Three in Calendrical Scrolls from Qumran," *Zeitschrift für Assyriologie* 95: 104-120.

Bergsma, John S. 2009. "The Relationship between Jubilees and the Early Enochic Books (*Astronomical Book* and *Book of the Watchers*)", in *Enoch and the Mosaic Torah: The Evidence of Jubilees*. Eds. G. Boccaccini & G. Ibba. Grand Rapids, Mi.: Eerdmans, 36-51.

Boccaccini, Gabriele & John J. Collins (eds.) 2007. *The Early Enoch Literature*. JSJSup 121, Leiden: Brill.

Böttrich, Christfried 1997. "Astrologie in der Henochtradition," *Zeitschrift für die alttestamentliche Wissenschaft* 109: 222-245.

Boustan, Ra'anan S. 2011. "Rabbinization and the Making of Early Jewish Mysticism," *Jewish Quarterly Review* 101: 482–501.

Brennan, Andrew 2004. "The Birth of Modern Science: Culture, Mentalities and Scientific Innovation," *Studies in History and Philosophy of Science* 35: 199–225.

Brooke, John Hendley 1991. *Science and Religion: Some Historical Perspectives*. Cambridge: Cambridge University Press.

Campbell-Kelly, Martin et al. 2003. *The History of Mathematical Tables: From Sumer to Spreadsheets*. Oxford and New-York: Oxford University Press.

Chambers, David Wade and Richard Gillespie. 2000. "Locality in the History of Science: Colonial Science, Technoscience, and Indigenous Knowledge," *Osiris* 15: 221-240.

Chemla, Karine (ed.) 2004. *History of Science, History of Text*. Boston Studies in the Philosophy of Science 238. Dordrecht: Springer.

Chesnutt, Randall D. 2010. "Oxyrhynchus Papyrus 2069 and the Compositional History of 1 Enoch," *Journal of Biblical Literature* 129: 485-505.

Coblentz Bautch, Kelley 2003. *A Study of the Geography of 1 Enoch 17-19: "No One Has Seen What I Have Seen."* Leiden, Brill.

Collins, John J. 1977. "Cosmos and Salvation: Jewish Wisdom and Apocalyptic in the Hellenistic Age," *History of Religions* 17: 121-142.

Collins, John J.1998. *The Apocalyptic Imagination: An Introduction to Jewish Apocalyptic Literature*. Grand Rapids, Mi.: Eerdmans.

Collins, John J. 2005. "The Reinterpretation of Apocalyptic Traditions in the Wisdom of Solomon," in *Jewish Cult and Hellenistic Culture. Essays on the Jewish Encounter with Hellenism and Roman Rule.* JSJSup 100. Leiden: Brill, 143-158

Cumont, Franz V.M. 1912. *Astrology and Religion among the Greeks and Romans*. New-York: GP Putnam.

Cuomo, Serafina 2007. *Technology and Culture in Greek and Roman Antiquity.* Cambridge: Cambridge University Press.

Dear, Peter 2005. "What is the History of Science the History of? Early Modern Roots of the Ideology of Modern Science," *Isis* 96: 390-406.

Dear, Peter 2009. "The History of Science and the History of the Sciences: George Sarton, Isis, and the Two Cultures," *Isis* 100 (2009): 89-93.

Dimant, Devorah 1974. *"The Fallen Angels" in the Dead Sea Scrolls and in the Apocryphal and Pseuepigraphic Books Related to them*. Dissertation, Hebrew University of Jerusalem. (in Hebrew)

Dimant, Devorah 2009. "Time, Torah and Prophecy at Qumran," in *Religiöse Philosophie und philosophische Religion der frühen Kaiserzeit*. Eds. R. Hirsch-Luipold, H. Goergemanns & M. von Albrecht. Studien und Texte zu Antike und Christentum 51. Tübingen: Mohr Siebeck, 147-161.

Drawnel, Henryk 2004. *An Aramaic Wisdom Text from Qumran: A New Interpretation of the Levi Document*. JSJSup 86. Leiden: Brill.

Drawnel, Henryk 2006. "Priestly Education in the Aramaic Levi Document (Visions of Levi) and Aramaic Astronomical Book (4Q208–211)", *Revue de Qumran* 22: 547-574.

Drawnel, Henryk 2007. "Moon Computation in the Aramaic Astronomical Book," *Revue de Qumran* 23: 3-41.

Drawnel, Henryk 2009. "Some Notes on Scribal Craft and the Origins of the Enochic Literature," *Henoch* 31: 66-72.

Drawnel, Henryk 2010. "Between Akkadian *tupšarrutu* and Aramaic *spr*: Some Notes on the Social Context of the Early Enochic Literature," *Revue de Qumran* 24: 373-403.

Drawnel, Henryk 2011. *The Aramaic Astronomical Book from Qumran: Text, Translation, and Commentary*. Oxford and New-York: Oxford University Press.

Eilberg-Schwartz, Howard 1987. "Creation and Classification in Judaism: From Priestly to Rabbinic Conceptions," *History of Religions* 26: 357-381.

Elior, Rachel 2004. *The Three Temples: On the Emergence of Jewish Mysticism*. Oxford: Littman Library.

Elshakry, Marwa 2010. "When Science Became Western: Historiographical Reflections," *Isis* 101: 98-109.

Farioli, Marcella 2010. "The Genesis of the Cosmos, the Search for *Arche* and the Finding of *Aitia* in Classical Greek Culture," in *Origins as a Paradigm in the Sciences and Humanities*. Ed. Paola Spinozzi and Alessandro Zironi. Göttingen: V&R Unipress, 195-209.

Feldman, William M. 1978. *Rabbinical Mathematics and Astronomy*. New-York: Hermon Press.

Finkel, Joshua 1928. "Review: Sarton on the History of Science," *Jewish Quarterly Review* 18: 445-448.

Fisch, Menachem 1997. *Rational Rabbis: Science and Talmudic Culture*. Bloomington: Indiana UP.

Fonrobert, Charlotte 2006. "The Semiotics of the Sexed Body in Early Halakhic Discourse," in *How Should Rabbinic Literature Be Read in the Modern World?*. Ed. M. Kraus. Piscataway, NJ: Gorgias, 79-104.

Fröhlich, Ida 2011. "Medicine and Magic in Genesis Apocryphon. Ideas on Human Conception and its Hindrances," *Revue de Qumran* 25: 177-198.

Funkenstein, Amos 1986. *Theology and the Scientific Imagination from the Middle Ages to the Seventeenth Century*. Princeton: Princeton University Press.

Galison, Peter 2004. "Removing Knowledge," *Critical Inquiry* 31:229-243.

Ganz, Solomon 1970. *Studies in Hebrew Astronomy and Mathematics*. New York: Ktav.

García-Martínez, Florentino 2010. "Beyond the Sectarian Divide: The 'Voice of the Teacher' as an Authority-Conferring Strategy in Some Qumran Texts," in *The Dead Sea Scrolls: Transmission of Traditions and Production of Texts*. Ed. S. Metso, H. Najman and E. Schuller. STDJ 92. Leiden: Brill, 227-244.

Geller, Markham 1991. "Akkadian Medicine in the Babylonian Talmud," in *A Traditional Quest, Essays in Honour of Louis Jacobs*, ed. D. Cohn-Sherbok. Sheffield: Academic Press, 102-112.

Geller, Markham 1995. "Babylonian Influence on Hellenistic Judaism," in *Civilizations of the Ancient Near East*. Ed. J.M. Sasson. New York: Scribner's, vol. I, 43-54.

Geller, Markham 1997. "The Last Wedge," *Zeitschrift für Assyriologie* 87: 43-95.

Geller, Markham 1998. "New Documents from the Dead Sea: Babylonian Science in Aramaic," in *Boundaries of the Ancient Near Eastern World: A Tribute to Cyrus H. Gordon*. Ed. M. Lubetski, C. Gottlieb, and S. Keller. JSOTSup 273. Sheffield: Academic Press, 224-229.

Geller, Markham 2000a. "The Survival of Babylonian Wissenschaft in Later Tradition," in *Melammu Symposia* I, ed. S. Aro and R. M. Whiting. Helsinki: Helsinki University Press, 1-6.

Geller, Markham 2000b. "An Akkadian Vademecum in the Babylonian Talmud," in *From Athens to Jerusalem, Medicine in Hellenized Jewish Lore and in Early Christian Literature*. Ed. S. Kottek, M. Horstmanshoff, G. Baader, and G. Ferngren. Rotterdam: Erasmus, 13-32.

Geller, Markham 2004. "Akkadian Healing Therapies in the Babylonian Talmud," *Max-Planck-Institut für Wissenschaftsgeschichte Preprint* 259: 1-60.

Glessmer, Uwe & Matthias Albani 1997. "An Astronomical Measuring Instrument from Qumran," in *The Provo International Conference on the Dead Sea Scrolls: Technological Innovations, New Texts, and Reformulated Issues*. Ed. D.W. Parry & E. Ulrich. STDJ 30. Leiden: Brill, 407-422.

Goff, Matthew J. 2007. *Discerning Wisdom: The Sapiential Literature of the Dead Sea Scrolls*. VTSup 116. Leiden: Brill.

Goodnick-Westenholz, Joan 1998. "Thoughts on Esoteric Knowledge and Secret Lore," in *Intellectual Life of the Ancient Near East: Papers Presented at the 43rd Rencontre Assyriologique Internationale, Prague, July 1-5, 1996*. Prague: Oriental Institute, 451-462.

Goshen-Gottstein, Alon 1995. "Is *Ma'aseh Bereshit* Part of Ancient Jewish Mysticism?" *Journal of Jewish Thought and Philosophy* 4: 185–201.

Grabbe, Lester L. 1995. *Priests, Prophets, Diviners, Sages: A Socio-historical Study of Religious Specialists in Ancient Israel*. Valley Forge, Pa.: Trinity Press International.

Graf, Fritz 1999. "Mythical Production: Aspects of Myth and Technology in Antiquity," in *From Myth to Reason? Studies in the Development of Greek Thought*. Ed. Richard Buxton. Oxford: Oxford University Press, 317-328.

Greenfield, Jonas C., Sokoloff, Michael & David Pingree 2000. "318. 4QZodiology and Brontology ar," in *Qumran Cave 4 XXVI Cryptic Texts and Miscellanea, Part 1*. Ed. S.J. Pfann et al. DJD 36. Oxford: Clarendon Press, 259-275.

Gruenwald, I. 1970–1971. "New Fragments from the Physiognomic and Chiromantic Literature" (Hebrew), *Tarbiz* 40: 301–319.

Gutas, Dimitri 1998. *Greek Thought, Arabic Culture: The Graeco-Arabic Translation Movement in Baghdad and Early 'Abbāsid Society (2nd-4th/8th-10th Centuries)*. London: Routledge.

Hartman, Lars 1979. *Asking for a Meaning. A Study of 1 Enoch 1-5*. CBNT 12. Lund: Gleerup.

Hengel, Martin 2000. "Qumran and Hellenism," in *Religion in the Dead Sea Scrolls*. Eds. J.J. Collins & R. Kugler. Grand Rapids, Mi.: Eerdmans, 46-56.

Henry, John 2002. *The Scientific Revolution and the Origins of Modern Science.* 2nd. ed. New York: Palgrave.

Himmelfarb, Martha 1993. *Ascent to Heaven in Jewish and Christian Apocalypses.* New York: Oxford University Press.

Himmelfarb, Martha 2006. "Merkavah Mysticism since Scholem: Rachel Elior's *The Three Temples*," in *Wege mystischer Gotteserfahrung: Judentum, Christentum und Islam*. Ed. P. Schäfer. Schriften des Historischen Kollegs Kolloquien 65. Oldenbourg: Oldenbourg Wissenschaftsverlag, 19-36.

Jacobus, Helen R. 2010. "4Q318: A Jewish Zodiac Calendar at Qumran?" in *The Dead Sea Scrolls: Texts and Context.* Ed. C. Hempel. STDJ 90. Leiden: Brill, 365-395.

Jaffee, Martin S. 2001. *Torah in the Mouth: Writing and Oral Tradition in Palestinian Judaism, 200 BCE-400 CE.* Oxford: Oxford University Press.

Jassen, Alex P. 2007. *Mediating the Divine: Prophecy and Revelation in the Dead Sea Scrolls and Second Temple Judaism*. STDJ 68. Leiden: Brill.

Jones, Alexander 1999. *Astronomical Papyri from Oxyrhynchus: (P. Oxy. 4133-4300a)*. Philadelphia: American Philosophical Society.

Jones, Alexander 2002. "Babylonian Lunar Theory in Roman Egypt: Two New Texts," in *Under One Sky, Astronomy and Mathematics in the Ancient Near East.* Ed. J. Steele and A. Imhausen. AOAT 297. Münster: Ugarit Verlag, 167-174.

Kessler, Gwynn 2009. *Conceiving Israel: the Fetus in Rabbinic Narratives.* Philadelphia: University of Pennsylvania Press.

Kiener, Ronald 1987. "Astrology in Jewish Mysticism from the Sefer Yezira to the Zohar," *Jerusalem Studies in Jewish Thought* 6 (3-4): 1*-42*.

Kiperwasser, Reuven 2009. "Three Partners in a Person: The Genesis and Development of Embryological Theory in Biblical and Rabbinic Judaism," *lectio difficilior* 2/2009 [http://www.lectio.unibe.ch]

Kister, Menahem 2004. "Wisdom Literature and its Relation to Other Genres," in *Sapiential Perspectives: Wisdom Literature in Light of the Dead Sea Scrolls: Proceedings of the Sixth International Symposium of the Orion Center for the Study of the Dead Sea Scrolls and Associated Literature, 20-22 May, 2001.* Ed. J.J. Collins, G.E. Sterling & R. Clements. STDJ 51. Leiden: Brill, 13-47.

Kleingünther, Adolf 1933. *PROTOS EURETES: Untersuchungen zur Geschichte einer Fragestellung*. Leipzig: Dietrich.

Klijn, Albertus F.J. 1977. *Seth in Jewish, Christian and Gnostic Literature.* NovTSup 46. Leiden: Brill.

Knibb, Michael A. 2003. "Which Parts of 1 Enoch Were Known to Jubilees? A Note on the Interpretation of Jubilees 4.16-25," in *Reading from Right to Left: Essays on the Hebrew Bible in Honour of David J.A. Clines.* Eds. J.C. Exum & H.G.M. Williamson. JSOTSup 373. Sheffield: Academic Press, 254-262.

Koch, Klaus 2007. "The Astral Laws as the Basis of Time, Universal History, and the Eschatological Turn in the Astronomical Book and the Animal Apocalypse of 1 Enoch," in *The Early Enoch Literature*. Eds. G. Boccaccini & J.J. Collins. JSJSup 121. Leiden: Brill, 119-137.

Kottek, Samuel S. 1996–1997. "Alexandrian Medicine in the Talmudic Corpus," *Koroth* 12: 85–87.

Kottek, Samuel S. 2006. "Medical Interest in Ancient Rabbinic Literature," in *The Literature of the Sages, Second Part* CRINT 3/2. Eds. Shmuel Safrai et al. Minneapolis: Fortress, 485-496.

Kuhn, Thomas S. 1996 [1962]. *The Structure of Scientific Revolutions*, 3rd ed. Chicago: University of Chicago Press.

Kvanvig, Helge S. 1988. *Roots of Apocalyptic: The Mesopotamian Background of the Enoch Figure and of the Son of Man*. Neukirchen-Vluyn: Neukirchener.

Kvanvig, Helge 2011. *Primeval History: Babylonian, Biblical, and Enochic. An Intertextual Reading*. JSJSup 149. Leiden: Brill.

Lange, Armin 1995. *Weisheit und Prädestination: Weisheitliche Urordnung und Prädestination in den Textfunden von Qumran*. STDJ 18. Leiden: Brill.

Langermann, Y. Tzvi 1997. "Science and the Kuzari," *Science in Context* 10: 495–522.

Langermann, Y. Tzvi 2002. "On the Beginnings of Hebrew Scientific Literature and on Studying History through 'maqbilot' (parallels)," *Aleph* 2: 169-189.

Latour, Bruno, and Steven Woolgar 1979. *Laboratory Life: the Social Construction of Scientific Facts.* Los Angeles: Sage.

Laudan, Larry 1983. "The Demise of the Demarcation Problem," in *Physics, Philosophy, and Psychoanalysis: Essays in Honor of Adolf Grünbaum* Boston Studies in the Philosophy of Science 76. Eds. R.S. Cohan and L. Laudan. Dordrecht: D. Reidel, 111–127.

Leicht, Reimund 2006. *Astrologumena Judaica: Untersuchungen zur Geschichte der astrologischen Literatur der Juden*. TSMJ 21. Tübingen: Mohr Siebeck.

Lenzi, Alan 2008. *Secrecy and the Gods: Secret Knowledge in Ancient Mesopotamia and Biblical Israel.* SAAS 19. Helsinki: Helsinki University Press.

Lenzi, Alan 2009. "Secrecy, Textual Legitimation, and Intercultural Polemics in the Book of Daniel," *Catholic Biblical Quarterly* 71: 330-348.

Lieber, Elinor 1984. "Asaf's Book of Medicines: A Hebrew Encyclopedia of Greek and Jewish Magic, Possibly Compiled on an Indian Model," *Dumbarton Oaks Papers* 38:233-249.

Liebes, Yehuda 2011. *The Cult of the Dawn. The Attitude of the Zohar towards Idolatry.* Jerusalem: Carmel (in Hebrew)

Lloyd, Geoffrey E. R. 2002. *The Ambitions of Curiosity: Understanding the World in Ancient Greece and China.* Cambridge: Cambridge University Press.

Long, Pamela O. 2001. *Openness, Secrecy, Authorship: Technical Arts and the Culture of Knowledge from Antiquity to the Renaissance.* Baltimore: Johns Hopkins University Press.

Lovejoy, Arthur O. and George Boas [1935] 1997. *Primitivism and Related Ideas in Antiquity*, Baltimore: Johns Hopkins University Press.

MacLeod, Roy 2000. "Introduction: Nature and Empire: Science and the Colonial Enterprise," *Osiris* 15: 1-13.

Martone, Corrado 2000. "Qumran and Stoicism: An Analysis of some Common Traits," in *The Dead Sea Scrolls Fifty Years after their Discovery 1947-1997*. Eds. L.H. Schiffman, E. Tov & J.C. VanderKam. Jerusalem: IES and the Shrine of the Book, 617-622.

McCants, William F. 2011. *Founding Gods, Inventing Nations: Conquest and Culture Myths from Antiquity to Islam*. Princeton: Princeton University Press.

Melamed, Abraham 2010. *The Myth of the Jewish Origins of Science and Philosophy.* Jerusalem: Magnes. [Hebrew]

Milik, Joseph T. 1976. *The Books of Enoch: Aramaic Fragments of Qumran Cave 4.* Oxford: Clarendon.

Momigliano, Arnaldo 1975. *Alien Wisdom: The Limits of Hellenization.* Cambridge: Cambridge University Press.

Morgenstern, Matthew 2000. "The Meaning of *byt mwldym* in the Qumran Wisdom Texts," *Journal of Jewish Studies* 51: 141-144.

Muntner, S. 1958. "Rufus of Samaria," *Israel Medical Journal* 17: 273–275.

Needham, Joseph et al. 1954-2004. *Science and Civilisation in China*, 7 vols. Cambridge: Cambridge University Press.

Neugebauer, Otto 1949. "The Astronomy of Maimonides and its Sources," *Hebrew Union College Annual* 22: 321-263.

Neugebauer, Otto 1985. "The 'Astronomical Chapters' of the Ethiopic Book of Enoch (72 to 82)," in Matthew Black, *The Book of Enoch or 1 Enoch: A New English Edition*. Leiden: Brill, 386-414.

Neugebauer, Otto and Henry B. van Hoesen 1959. *Greek Horoscopes*. MAPS. Philadelphia: American Philosophical Society.

Neusner, Jacob 1988. "Why No Science in Judaism?" *Shofar* 6.3: 45–71.

Neusner, Jacob 1989. "Science and Magic, Miracle and Magic, in Formative Judaism: The System and the Difference," in *Religion, Science, and Magic in Concert and in Conflict*. Ed. J. Neusner, E. S. Frerichs, P. V. M. Flesher. New York: Oxford University Press, 61-81.

Newsom, Carol A. 1980. "The Development of Enoch 6-19: Cosmology and Judgment," *Catholic Biblical Quarterly* 42: 310-329.

Nickelsburg, George W.E. 1977. "Apocalyptic and Myth in 1 Enoch 6-11," *Journal of Biblical Literature* 96: 383-405.

Nickelsburg, George W.E. 2001. *1 Enoch 1: A Commentary on the Book of 1 Enoch*. Hermeneia. Minnesota: Fortress.

Nickelsburg, George W.E, and James C. VanderKam 2012. *1 Enoch 2: A Commentary on the Book of 1 Enoch: Chapters 37-82*. Hermeneia. Minneapolis: Fortress.

Niehoff, Maren 2001. "Inscribing Jewish Culture into Nature," *Philo on Jewish Identity and Culture*. TSAJ 86. Tübingen: Mohr Siebeck, 247-266.

Orlov, Andrei A. 2001. "Overshadowed by Enoch's Greatness: 'Two Tablets' Traditions from the *Book of Giants* to Paleia Historica," *Journal for the Study of Judaism* 32: 137-158.

Parpola, Simo 1993. "Mesopotamian Astrology and Astronomy as Domains of the Mesopotamian 'Wisdom'," in *Die Rolle der Astronomie in den Kulturen Mesopotamiens*. Grazer morgenlaendlische Studien 3. ed. H.D. Galter. Graz: GrazKult.

Patai, Raphael 1994. *The Jewish Alchemists: A History and Source Book*. Princeton: Princeton University Press.

Pfaff, Franz 1932. "Rufus aus Samaria," *Hermes* 67: 356–359.

Pfann, Stephen J. 2000. "The Writings in Esoteric Script from Qumran," in *The Dead Sea Scrolls Fifty Years after their Discovery*. Eds. L.H. Schiffman & E. Tov. Jerusalem: IES and The Shrine of the Book, 177-190.

Pfann, Stephen J. 2001. *The Character of the Early Essene Movement in the Light of the Manuscripts Written in Esoteric Scripts from Qumran*. Dissertation, Hebrew University of Jerusalem.

Pines, Shlomo 1975. "The Oath of Asaph the Physician and Yohanan Ben Zabda," *Proceedings of the Israel Academy of Sciences and Humanities* 9: 223–264.

Popović, Mladen 2006. "Physiognomic Knowledge in Qumran and Babylonia: Form, Interdisciplinarity, and Secrecy," *Dead Sea Discoveries* 13: 150-176.

Popović, Mladen 2007. *Reading the Human Body: Physiognomics and Astrology in the Dead Sea Scrolls and Hellenistic-Early Roman Period Judaism*. STDJ 67. Leiden: Brill.

Popović, Mladen 2010. "The Emergence of Aramaic and Hebrew Scholarly Texts: Transmission and Translation of Alien Wisdom," in *The Dead Sea Scrolls: Transmission of Tradition and Production of Texts*. Ed. S. Metso, H. Najman & E. Schuller. STDJ 92. Leiden: Brill, 81-114.

Preuss, Julius 1978 [1911]. *Biblical and Talmudic Medicine*. Trans. F. Rosner. New York: Sanhedrin Press.

Principe, Lawrence M. 2011. "Alchemy Restored," *Isis* 102: 305-312.

Puech, Émile 2009. *Qumrân Grotte 4 XXVII. Textes Araméens. Deuxième partie. 4Q550-4Q575a. 4Q580-4Q587 et appendices*. DJD 37. Oxford: Clarendon

Pyenson, Lewis 1993. "The Ideology of Western Rationality: History of Science and the European Civilizing Mission," *Science & Education* 2: 329-343.

Rand, Michael 2009. "Clouds, Rain, and the Upper Waters: From Bereshit Rabbah to the Piyyuṭim of Eleazar be-rabbi Qillir," *Aleph* 9: 13-39.

Reed, Annette Yoshiko (forthcoming). "Enoch, Eden, and the Beginnings of Jewish Cosmography," in *The Cosmography of Paradise*. Ed. Alessandro Scafi. London-Turin: The Warburg Institute-Nino Aragno.

Reed, Annette Yoshiko 2004. "Abraham as Chaldean Scientist and Father of the Jews: Josephus, *Ant.* 1.154–168, and the Greco-Roman Discourse about Astronomy/Astrology," *Journal for the Study of Judaism* 35:119-158.

Reed, Annette Yoshiko 2005. *Fallen Angels and the History of Judaism and Christianity: The Reception of Enochic Literature*. Cambridge: Cambridge University Press.

Reed, Annette Yoshiko 2007. "Was there Science in Ancient Judaism? Historical and Cross-cultural Reflections on 'Religion' and 'Science'," *Studies in Religion/Sciences Religieuse* 36: 461–495.

Reed, Annette Yoshiko 2008a. "The Origins of the *Book of the Watchers* as 'Apocalypse' and its Reception as 'Apocrypha'," *Henoch* 30: 55–59.

Reed, Annette Yoshiko 2008b. "Asaph ha-Rofe," "Domnus," "Gamaliel VI," "Rufus of Samaria," "Samuel of Nehardea," "Theodos of Alexandria," and "Zakhalias of Babylon," in *Encyclopedia of Ancient Natural Scientists: The Greek Tradition and its Many Heirs*. Ed. Paul T. Keyser and Georgia Irby-Massie. London: Routledge, 168, 275, 342-3, 721, 726, 788-789, and 843.

Reed, Annette Yoshiko 2010. "From 'Pre-Emptive Exegesis' to 'Pre-Emptive Speculation'? *Ma'aseh Bereshit* in *Bereshit Rabbah* and *Pirqei deRabbi Eliezer*," in *With Letters of Light - Otiyot Shel Or: Studies in Early Jewish Apocalypticism and Mysticism in Honour of Rachel Elior*. Ed. D. Arbel and A. Orlov. Berlin: De Gruyter, 115-132.

Regev, Eyal 2007. *Sectarianism in Qumran: A Cross-Cultural Perspective*. Berlin: de Gruyter.

Rey, Jean-Sébastian 2009. *4QInstruction: sagesse et eschatology*. STDJ 81. Leiden: Brill.

Roberts, Lissa 2009. "Situating Science in Global History: Local Exchanges and Networks of Circulation," *Itinerario* 33: 9-30.

Rochberg-Halton, Francesca 1989. "Babylonian Seasonal Hours," *Centaurus* 32: 146-170.

Rochberg, Francesca 1998. *Babylonian Horoscopes*. TAPS 88. Philadelphia: American Philosophical Society.

Rochberg, Francesca 2002. "A Consideration of Babylonian Astronomy within the Historiography of Science," *Studies in the History and Philosophy of Science Part A*, 33: 661-684.

Rochberg, Francesca 2004. *The Heavenly Writing: Divination, Horoscopy, and Astronomy in Mesopotamian Culture*. Cambridge: Cambridge University Press.

Ruderman, David B. 1995. *Jewish Thought and Scientific Discovery in Early Modern Europe*. New Haven and London: Yale University Press.

Sabra, Abdelhamid I. 1987. "The Appropriation and Subsequent Naturalization of Greek Science in Medieval Islam," *History of Science* 25: 225-243.

Sacchi, Paolo 2008. "The *Book of the Watchers* as an Apocalyptic and Apocryphal Text," *Henoch* 30: 9-26.

Safrai, Zeev 2006. "Geography and Cosmography in Talmudic Literature," in *The Literature of the Sages*. CRINT 3/2. Ed. Shmuel Safrai, et al. Minneapolis: Fortress, 497-508.

Sanders, Seth L. 2009. "The First Tour of Hell," *Journal of Ancient Near Eastern Religions* 9:151-170.

Sanders, Seth L. 2012. "Apocalyptic Science," in *AJS Perspectives*, Fall 2012: 16-17.

Sarton, George 1929. *Introduction to the History of Science*, vol. 1: *From Homer to Omar Khayyam*. Baltimore: Williams and Wilkins.

Schäfer, Peter 1988. "Ein neues Fragment zur Metoposkopie und Chiromantik," in *Hekhalot-Studien*. Tübingen: Mohr Siebeck, 84-95.

Schäfer, Peter 2004. "In Heaven as it is in Hell: The Cosmology of *Seder Rabbah de-Bereshit*," in *Heavenly Realms and Earthly Realities in Late Antique Religions*. Eds. Ra'anan S. Boustan and Annette Yoshiko Reed. Cambridge: Cambridge University Press, 233-274.

Schäfer, Peter 2005. "From Cosmology to Theology: The Rabbinic Appropriation of Apocalyptic Cosmology," in *Creation and Re-Creation in Jewish Thought: Festschrift in Honor of Joseph Dan on the Occasion of His Seventieth Birthday*. Ed. Rachel Elior and Peter Schäfer. Tübingen: Mohr Siebeck, 39-58.

Schäfer, Peter 2008. "Bereshit bara Elohim: Bereshit Rabba, Parasha 1, Reconsidered," in *Empsychoi Logoi: Religious Innovations in Antiquity: Studies in Honour of Pieter Willem van der Horst.* Ed. A. Houtman, A. de Jong, and M. Misset-van de Weg. Leiden: Brill, 267-289.

Schaffer, Simon and Steven Shapin 1985. *Leviathan and the Air-Pump: Hobbes, Boyle, and the Experimental Life*. Princeton: Princeton University Press.

Schmid, Hans H. 1968. *Gerechtigkeit als Weltordnung: Hintergrund und Geschichte der alttestamentlichen Gerechtigkeitsbegriffes*. Tübingen: Mohr Siebeck.

Schmidt, Francis 1997. "Astrologie juive ancienne: Essai d'interprétation de 4QCryptique (4Q186)," *Revue de Qumran* 18: 125-141.

Schnabel, Paul 1923. *Berossos und die babylonisch-hellenistische Literatur*. Leipzig: Teubner.

Scurlock, JoAnn 2006. "Sorcery in the Stars: STT 300, BRM 4.19-20 and the Mandaic Book of the Zodiac," *Archiv für Orientforschung* 51: 125-146.

Scurlock, JoAnn and Farouq N.H. Al-Rawi 2006, "A Weakness for Hellenism," in *If a Man Builds a Joyful House. Assyriological Studies in Honor of Erle Verdun Leichty*. Ed. A.K. Guinan. Cuneiform Monographs 31. Leiden: Brill, 357-382.

Séd, Nicolas 1981. *La mystique cosmologique juive*. Paris: Editions de l'Ecole des hautes études en sciences socials.

Segal, Michael 2007. *The Book of Jubilees: Rewritten Bible, Redaction, Ideology and Theology*. JSJSup 117. Leiden: Brill.

Sela, Shlomo 2003. *Abraham Ibn Ezra and the Rise of Medieval Hebrew Science*. Brill's Series in Jewish Studies 32. Leiden: Brill.

Shemesh, Abraham Ofir 2006. "Biology in Rabbinic Literature: Fact and Folklore," in *The Literature of the Sages*. CRINT 3/2. Ed. Shmuel Safrai, et al. Minneapolis: Fortress, 509-519.

Shemesh, Abraham Ofir 2010. "Therapeutic Bathing in Rabbinic Literature: Halachic Issues and their Background in History and Realia," *Jewish Medical Ethics* 7: 57–67.

Stern, Sacha 1996. "Fictitious Calendars: Early Rabbinic Notions of Time, Astronomy, and Reality," *Jewish Quarterly Review* 87:103-129.

Stern, Sacha 2005. "Rachel Elior on Ancient Jewish Calendars: A Critique," *Aleph* 5: 287-292.

Stone, Michael E. 1976. "Lists of Revealed Things in the Apocalyptic Literature," in *Magnalia Dei - the Mighty Acts of God. Essays on the Bible and Archaeology in Memory of G. Ernest Wright*. Eds. F.M. Cross, W.E. Lemke & P.D. Miller. Garden City, NY: Doubleday, 414-452.

Stone, Michael E. 1987. "The Parabolic Use of Natural Order in Judaism of the Second Temple Age," in *Gilgul. Festschrift R.J.Z. Wervlowsky*. Numen Sup. 50. Eds. S. Shaked, D. Shulman & G.G. Stroumsa. Leiden: Brill, 298-308.

Stone, Michael E. 1988. "Enoch, Aramaic Levi and Sectarian Origins," *Journal for the Study of Judaism* 19: 159-170.

Talmon, Shemaryahu, Ben-Dov, Jonathan, & Uwe Glessmer 2001. *Qumran Cave 4 XVI. Calendrical* texts. DJD 21. Oxford: Clarendon Press.

Tambiah, Stanley 1990. *Magic, Science, Religion, and the Scope of Rationality*. Cambridge: Cambridge University Press.

Tavardon, Paul 2010. *Le disque de Qumrân*. CRB 75. Paris: Gabalda.

Taylor, Joan E. 2009. "'Roots, Remedies and Properties of Stones': The Essenes, Qumran and Dead Sea Pharmacology," *Journal of Jewish Studies* 60: 226-244.

Thomas, Samuel I. 2009. *The "Mysteries" of Qumran: Mystery, Secrecy, and Esotericism in the Dead Sea Scrolls*. Atlanta: SBL.

Thrade, K. 1962. "Erfinder II," in *Reallexikon für Antike und Christentum*. Ed. T. Klausner. Vol. V. Stuttgart: Anton Hiersemann, vol. 5, 1191-1278.

Tigchelaar, Eibert J.C. 2001. *To Increase Learning for the Understanding Ones: Reading and Reconstructing the Fragmentary Early Jewish Sapiential Text 4QInstruction*. STDJ 44. Leiden: Brill.

Tigchelaar, Eibert J.C. 2010. "Aramaic Texts from Qumran and the Authoritativeness of Hebrew Scriptures: Preliminary Observations," in *Authoritative Scriptures in Ancient Judaism*. Ed. M. Popović. JSJSup 141. Leiden: Brill, 155-171.

Van Bladel, Kevin T. 2009. *The Arabic Hermes: from Pagan Sage to Prophet of Science*. Oxford and New-York: Oxford University Press.

van Kooten, George H. 1999. "Enoch, the 'Watchers', Seth's Descendants and Abraham as Astronomer. Jewish Applications of the Greek Motif of the First Inventor (300 BCE - CE 100)", in *Recycling Biblical Figures: Papers Read at a NOSTER Colloquium in Amsterdam 12-13 May 1997*. Eds. J.W. van Henten & A. Brenner. Leiderdorp: Deo Publishing, 292-316.

Van Seters, John 1992. *Prologue to History: The Yahwist as Historian in Genesis*. Westminster: John Knox Press.

VanderKam, James C. 1983. *Enoch and the Growth of an Apocalyptic Tradition*. CBQMS 16. Washington D.C.: Catholic Biblical Association of America.

VanderKam, James C. 2002. *From Revelation to Canon: Studies in the Hebrew Bible and Second Temple Literature*. Leiden: Brill.

VanderKam, James C. 2007. "Mapping Second Temple Judaism," in *The Early Enoch Literature*. Ed. Gabriele Boccaccini and John J. Collins. JSJSup 121. Leiden: Brill, 1-20.

Veltri, Giuseppe 1998. "The Rabbis and Pliny the Elder: Jewish and Greco-Roman Attitudes toward Magic and Empirical Knowledge," *Poetics Today* 19: 63-89.

Verbrugghe, Gerald & John M. Wickersham 1996. *Berossos and Manetho, Introduced and Translated: Native Traditions in Ancient Mesopotamia and Egypt*. Ann Arbor: University of Michigan Press.

Walbridge, John 1998. "Explaining Away the Greek Gods in Islam," *Journal of the History of Ideas* 59: 389–403.

Wright, Archie T. 2005. *The Origin of Evil Spirits: The Reception of Genesis 6.1-4 in Early Jewish Literature*. WUNT n.s. 198. Tübingen: Mohr Siebeck.

Wright, Benjamin G. 2007. "1 Enoch and Ben-Sirah: Wisdom and Apocalypticism in Relationship," in *The Early Enoch Literature*. Eds. G. Boccaccini & J.J. Collins. JSJSup 121. Leiden: Brill, 159-176.

Zhmud, Leonid 2006. *The Origin of the History of Science in Classical Antiquity*. Trans. Alexander Chernoglazov. Berlin: de Gruyter.

Index